Bastian Jungnitsch

Criteria for genuine multiparticle quantum correlations

Bastian Jungnitsch

Criteria for genuine multiparticle quantum correlations

Methods of entanglement detection and verification

Südwestdeutscher Verlag für Hochschulschriften

Impressum/Imprint (nur für Deutschland/only for Germany)
Bibliografische Information der Deutschen Nationalbibliothek: Die Deutsche Nationalbibliothek verzeichnet diese Publikation in der Deutschen Nationalbibliografie; detaillierte bibliografische Daten sind im Internet über http://dnb.d-nb.de abrufbar.
Alle in diesem Buch genannten Marken und Produktnamen unterliegen warenzeichen-, marken- oder patentrechtlichem Schutz bzw. sind Warenzeichen oder eingetragene Warenzeichen der jeweiligen Inhaber. Die Wiedergabe von Marken, Produktnamen, Gebrauchsnamen, Handelsnamen, Warenbezeichnungen u.s.w. in diesem Werk berechtigt auch ohne besondere Kennzeichnung nicht zu der Annahme, dass solche Namen im Sinne der Warenzeichen- und Markenschutzgesetzgebung als frei zu betrachten wären und daher von jedermann benutzt werden dürften.

Coverbild: www.ingimage.com

Verlag: Südwestdeutscher Verlag für Hochschulschriften GmbH & Co. KG
Heinrich-Böcking-Str. 6-8, 66121 Saarbrücken, Deutschland
Telefon +49 681 37 20 271-1, Telefax +49 681 37 20 271-0
Email: info@svh-verlag.de

Approved by: Innsbruck, Leopold-Franzens-Universität, Diss., 2012

Herstellung in Deutschland:
Schaltungsdienst Lange o.H.G., Berlin
Books on Demand GmbH, Norderstedt
Reha GmbH, Saarbrücken
Amazon Distribution GmbH, Leipzig
ISBN: 978-3-8381-3188-7

Imprint (only for USA, GB)
Bibliographic information published by the Deutsche Nationalbibliothek: The Deutsche Nationalbibliothek lists this publication in the Deutsche Nationalbibliografie; detailed bibliographic data are available in the Internet at http://dnb.d-nb.de.
Any brand names and product names mentioned in this book are subject to trademark, brand or patent protection and are trademarks or registered trademarks of their respective holders. The use of brand names, product names, common names, trade names, product descriptions etc. even without a particular marking in this works is in no way to be construed to mean that such names may be regarded as unrestricted in respect of trademark and brand protection legislation and could thus be used by anyone.

Cover image: www.ingimage.com

Publisher: Südwestdeutscher Verlag für Hochschulschriften GmbH & Co. KG
Heinrich-Böcking-Str. 6-8, 66121 Saarbrücken, Germany
Phone +49 681 37 20 271-1, Fax +49 681 37 20 271-0
Email: info@svh-verlag.de

Printed in the U.S.A.
Printed in the U.K. by (see last page)
ISBN: 978-3-8381-3188-7

Copyright © 2012 by the author and Südwestdeutscher Verlag für Hochschulschriften GmbH & Co. KG and licensors
All rights reserved. Saarbrücken 2012

Abstract

This thesis is devoted to different aspects of the detection and characterization of quantum correlations in multiparticle systems. These include the statistical verification of non-locality and entanglement in experiments, a versatile approach for the detection of genuine multipartite entanglement that will be applied to different classes of states and the characterization of entanglement using a multipartite hidden-variable theory.

We start by investigating statistical effects on the confidence with which one can ensure that a multipartite state is non-local and therefore entangled. It turns out that for the standard photonic error model, the statistical significance of a non-locality test with comparably low violation, the so-called Mermin inequality, can exceed the significance of a test with high violation, namely of the Ardehali inequality. We also report about an experiment with four photons that verifies this behavior. Moreover, we also find that the range of white noise in which the Mermin inequality achieves a higher statistical significance grows exponentially with an increasing number of particles.

Then, we pass on to the detection of genuine multipartite entanglement given the density matrix of a state. Using supersets of the sets of separable states, we introduce a criterion for genuine multipartite entanglement that can be implemented as a semidefinite program and test its performance on several example cases. Furthermore, this criterion naturally leads to an entanglement monotone that generalizes the bipartite negativity.

Subsequently, the criterion is applied to the class of graph states. In this way, we obtain analytical construction methods for entanglement criteria, so-called entanglement witnesses, for many different graph states and an arbitrary number of qubits. These witnesses turn out to be strong, as their white noise tolerance converges to the maximum possible value, namely one, for a growing number of qubits. At the same time, the additional experimental effort in terms of settings to be measured, stays constant.

Furthermore, as the criterion performs so well on graph states, we consider the question whether it can provide necessary and sufficient criteria for entanglement. We show that this is indeed the case for some special graph-diagonal states. In addition, our line of argument also provides deeper insights into the properties of our criterion and methods to construct biseparable graph-diagonal states.

Finally, we turn to the characterization of quantum mechanical correlations given by the bipartite non-local model introduced by A. Leggett. We present different ways of extending this model to the case of many particles and also derive an inequality that shows the imcompatibility of such multipartite Leggett models with quantum mechanics.

Contents

1	**Introduction**	**1**
2	**Setting the stage**	**3**
	2.1 Characterization of entanglement	3
	2.1.1 Bipartite entanglement	3
	2.1.2 Multipartite entanglement	4
	2.2 Detection of entanglement	5
	2.2.1 PPT criterion	6
	2.2.2 Entanglement witnesses	7
	2.3 Quantification of entanglement	10
	2.3.1 Bipartite entanglement measures	10
	2.3.2 Multipartite entanglement measures	13
	2.4 Hidden-variable theories	14
	2.4.1 Bipartite Bell inequalities	14
	2.4.2 Multipartite Bell inequalities	15
	2.4.3 Leggett inequalities	18
	2.5 Graph states	20
	2.5.1 Definition	20
	2.5.2 Graph state basis	22
	2.5.3 Local unitary operations on graph states	23
	2.6 Semidefinite programming	24
	2.6.1 General form of a semidefinite program	24
	2.6.2 Properties of a semidefinite program	26
3	**Bell inequalities: Statistical significances in experiments**	**29**
	3.1 Statement of the problem	30

3.2	Variance as error	30
3.3	Error model for multi-photon experiments	31
3.4	Bell inequalities for four particles	34
3.5	Experimental setup	43
3.6	Experimental results	44
3.7	Discussion	45

4 Entanglement detection via PPT mixtures   47
4.1	General idea	47
4.2	Characterization via entanglement witnesses	49
4.3	Practical evaluation of the criterion	49
4.4	Numerical examples	53
4.5	An analytical witness for the W state	57
4.6	Discussion	58

5 Entanglement witnesses for graph states   61
5.1	Graph-diagonal states		62
5.2	Fully decomposable witnesses		63
	5.2.1	Graph states up to 6 qubits	63
	5.2.2	Analytical construction methods	65
	5.2.3	Extended construction method	68
5.3	Fully PPT witnesses		70
	5.3.1	Arbitrary graph states	70
	5.3.2	Extended construction method	71
	5.3.3	2D cluster state	74
5.4	Entanglement monotone		75
5.5	Conclusion		76
5.6	Proofs		76
5.7	Witnesses		103

6 Necessary and sufficient criteria for graph state entanglement   107
6.1	Cluster-diagonal states of four qubits	108

	6.2	Five-qubit graph states	113

- 6.2 Five-qubit graph states 113
  - 6.2.1 The state Y_5 113
  - 6.2.2 The linear cluster state Cl_5 114
  - 6.2.3 The ring cluster state R_5 115
- 6.3 Connection with the theory of PPT mixtures 117
- 6.4 PPT mixtures and the five-qubit Y_5 state 118
- 6.5 Generalizations to more than five particles 120
  - 6.5.1 A generalization to Y_N-states 120
  - 6.5.2 Biseparable decompositions for linear cluster states 122
- 6.6 Conclusion 122

7 Multipartite Leggett models   123
- 7.1 Leggett's assumption on one qubit 123
  - 7.1.1 An inequality for hybrid Leggett models with assumptions on a single party . 125
- 7.2 An assumption on two qubits 128
- 7.3 Conclusion 130

8 Conclusion   131

1 Introduction

In the early years of quantum mechanics, physicists felt that there was something counterintuitive about quantum mechanics. This skepticism was phrased, most famously, by Einstein, Podolsky and Rosen [1], who believed that quantum mechanics is not a complete description of reality. In 1964, Bell showed that quantum mechanics possesses properties substantially different from classical theories by proving that such intuitively expected properties as locality and realism are incompatible with quantum mechanics [2]. These incompatibilities arise from the phenomenon of entanglement which was first described by Schrödinger in 1935 [3]. However, it was not until 1989, when the definition of entanglement was made precise [4].

At that time, the field of quantum information started to develop and a second aspect of entanglement came into focus: It can be used as a resource which permits tasks like entanglement-based quantum cryptography [5], teleportation [6] and measurement-based quantum computation [7,8]. The evolution of quantum information was boosted further when quantum algorithms were discovered that could perform certain tasks faster than any classical algorithm known. Shor's algorithm is exponentially faster than the best known classical algorithm for prime factorization [9], while Grover's algorithm outperforms classical algorithms for search in an unsorted database [10].

However, the connection of entanglement and a computational speed-up is not yet clear, and only few quantitative statements can be made. For example, if, in a quantum algorithm using only pure states, the maximal amount of entanglement present during the computation grows at most logarithmically in the number of qubits, the algorithm can be simulated classically using only a polynomial amount of time and memory [11]. It should also be mentioned that quantum discord [12], another measure of correlations, is discussed as a resource in some specialized quantum computational tasks involving totally mixed states [13]. Nevertheless, entanglement is necessary for teleportation, for the violation of a Bell inequality and plays a crucial role in measurement-based quantum computation and quantum metrology [14].

Due to its central role, entanglement has been under intense research in the last two decades. On one hand, its theoretical characterization received a lot of attention [15–19]. It turned out that, compared to a system of two particles, the characterization of entanglement for three or more particles is far more complex [20–23].

On the other hand, experimentalists have made substantial progress and are nowadays able to manipulate up to 14 qubits in experiments with trapped ions [24] and 10 qubits in photonic experiments [25]. Also, experiments using super-conducting qubits are making considerable progress [26,27].

In this thesis, we consider three different facets of quantum correlations in multipartite systems: We analyze their detection in experiment, develop tools for their theoretical verification and investigate their characterization.

For the first aspect, we will ask the question how sure one can be that a state in experiment is really entangled. This question will be tackled at the example of multipartite Bell inequalities whose violation — and thus also the fact that the used state is entangled — is to be confirmed in a photonic experiment. To this end, one needs to take into account statistical effects. We find that these statistical effects play an important role and that different Bell inequalities can exhibit very different behavior when one tries to confirm the presence of quantum correlations in experiment with large statistical significance.

Moreover, we will then proceed to the task of determining whether a multipartite state is genuinely multipartite entangled given its density matrix. A large part of this thesis is devoted to introducing and applying a criterion that enables one to both detect and quantify multipartite entanglement and is strongly connected to a certain class of witnesses. We evaluate the criterion numerically and illustrate its main properties, such as the fact that it detects any three-qubit permutation invariant state, can be applied in the case of incomplete information and provides a way to quantify entanglement. Then, we apply it to a certain class of states, namely graph states, for which we find analytical construction methods for tools that detect entanglement, so-called entanglement witnesses. As a given entanglement witness only supplies a sufficient criterion for entanglement, we will then proceed to the question whether our approach helps to find necessary and sufficient criteria for entanglement. We present some classes of graph diagonal states in which this is the case.

Finally, we turn to the characterization of quantum correlations and extend a class of hidden-variable models, namely Leggett models, from the bipartite to the multipartite case [28]. Instead of the assumption of locality as in a Bell inequality, Leggett inequalities basically assume that locally, all expectation values seem to come from pure states. We provide an inequality that holds for a certain class of the presented multipartite Leggett models. Then, we show that quantum mechanics violates this inequality and therefore the assumptions of the considered hidden-variable model.

This thesis is structured as follows: First, in Sec. 2, we lay out the basic notions and definitions used in this work. In Sec. 3, we investigate the statistical behavior of multipartite Bell inequalities in photonic experiments. Section 4 then introduces a criterion and a monotone for genuine multipartite entanglement and illustrates the idea behind the criterion and its main properties. Then, we analytically apply this criterion the class of graph states in Sec. 5. Section 6 then asks whether our approach leads to necessary and sufficient conditions for entanglement in graph states and presents cases in which it does. Finally, Sec. 7 covers the generalization of Leggett models to the multipartite case. Section 8 then presents a summary of the results of this thesis.

In the following chapter, we will start by laying the foundations and presenting the definitions of the notions mentioned before.

2 Setting the stage

This chapter is devoted to the definitions which later will be used in this thesis.

2.1 Characterization of entanglement

In the first section, we will learn the definitions of entanglement in systems composed by two particles and in systems composed by a larger number of particles.

2.1.1 Bipartite entanglement

Pure states

First, we consider two quantum mechanical systems A and B that are controlled by two persons named Alice and Bob, respectively. Let us start with the case of pure states. To each of the two systems we associate a Hilbert space, $\mathcal{H}_A$ and $\mathcal{H}_B$, respectively. Then, the state of the combined system is described by a normalized vector $|\psi\rangle \in \mathcal{H}$, where the Hilbert space $\mathcal{H}$ is given as the tensor product $\mathcal{H} = \mathcal{H}_A \otimes \mathcal{H}_B$. Moreover, $|\psi\rangle$ can be written as

$$|\psi\rangle = \sum_{i=1}^{d_A} \sum_{j=1}^{d_B} c_{ij} |a_i\rangle \otimes |b_j\rangle , \qquad (2.1)$$

where $|a_i\rangle \in \mathcal{H}_A$, $|b_j\rangle \in \mathcal{H}_B$ and d_A is the dimension of $\mathcal{H}_A$, d_B the dimension of $\mathcal{H}_B$. Furthermore, c_{ij} are complex coefficients. Note that, from now on, we will most often use the shorthand notations $|a_i\rangle|b_j\rangle$ or $|a_i\, b_j\rangle$ instead of $|a_i\rangle \otimes |b_j\rangle$.

Separability and entanglement is now defined in the following way for bipartite pure states.

Definition 1. *A pure state $|\psi\rangle \in \mathcal{H}_A \otimes \mathcal{H}_B$ is called **separable** if it can be written as*

$$|\psi\rangle = |a\rangle|b\rangle \qquad (2.2)$$

*for some states $|a\rangle \in \mathcal{H}_A$, $|b\rangle \in \mathcal{H}_B$. Otherwise, it is called **entangled**.*

Finally, a very useful tool for the characterization of entanglement in a bipartite setting is the so-called **Schmidt decomposition**.

Lemma 2. *Any pure state* $|\psi\rangle \in \mathcal{H}_A \otimes \mathcal{H}_B$ *can be written as*

$$|\psi\rangle = \sum_{i=1}^{r} \lambda_i |a_i\rangle |b_i\rangle, \qquad (2.3)$$

*where the so-called **Schmidt coefficients** λ_i are strictly positive and real. Moreover, they are unique and $\sum_{i=1}^{r} \lambda_i^2 = 1$. The number r of terms in the above sum is called **Schmidt rank** and obeys $r \leq \min\{d_A, d_B\}$, where d_A and d_B are the dimensions of $\mathcal{H}_A$ and $\mathcal{H}_B$.*

The proof of Lemma 2 can be found, e.g. in Ref. [29].

Mixed states

In general, the state of a quantum mechanical system is described by a positive, Hermitian linear operator of unit trace. This is due to the fact that a system can also be in a mixed state, i.e. its density matrix is given by

$$\varrho = \sum_i p_i |\phi_i\rangle \langle \phi_i|, \qquad (2.4)$$

where the p_i form a probability distribution and thus $p_i \geq 0$ and $\sum_i p_i = 1$. We can now define the notion of entanglement for mixed states.

Definition 3. *A state ϱ is called **separable** if there exist states ϱ_A^i for Alice and states ϱ_B^i for Bob, such that*

$$\varrho_{\text{sep}} = \sum_i p_i \, \varrho_A^i \otimes \varrho_B^i, \qquad (2.5)$$

*where $p_i \geq 0$ for all i and $\sum_i p_i = 1$. Otherwise, it is called **entangled**.*

A separable state can also be understood as a state that can be prepared via operations performed locally by Alice and Bob and classical communication between them. In particular, Alice can use a random number generator that outputs i with probability p_i. She can communicate her random output classically to Bob. Then, Alice prepares the state $|a_i\rangle$ locally, while Bob prepares $|b_i\rangle$. In this way, they can produce any separable state. Note that the set of local operations and classical communication is usually labelled by LOCC. We will come back to this kind of operations in Sec. 2.3.1.

2.1.2 Multipartite entanglement

In the case of three or more particles, the notion of entanglement becomes more complex. For illustration purposes, we consider three particles controlled by Alice, Bob and Charlie. The notions defined here can be generalized to a larger number of particles in a straightforward way. Let us first define a class of states that contain no entanglement between any particles.

2.2 Detection of entanglement

Definition 4. *A state ϱ is called **fully separable** if it can be written as*

$$\varrho_{\text{fs}} = \sum_i p_i \, \varrho_A^i \otimes \varrho_B^i \otimes \varrho_C^i \,, \tag{2.6}$$

where $p_i \geq 0$ for all i and $\sum_i p_i = 1$. A multipartite state ϱ that cannot be written in this form is called **entangled**.

There is, however, a kind of multipartite entanglement that is stronger than being not fully separable. In order to define this kind of entanglement, we need to consider bipartitions. For example, one can combine systems A and B and interpret them as a single system. In this way, one has defined a bipartition $AB|C$ and we are back in the bipartite situation, i.e. we can define states that are separable with respect to this bipartition according to Def. 3.

Definition 5. *A state ϱ is called **biseparable** if it can be written as*

$$\varrho_{\text{bs}} = p_1 \, \varrho_{AB|C}^{\text{sep}} + p_2 \, \varrho_{A|BC}^{\text{sep}} + p_3 \, \varrho_{B|AC}^{\text{sep}} \tag{2.7}$$

where $p_i \geq 0$ for all i and $\sum_{i=1}^{3} p_i = 1$. Also, $\varrho_{AB|C}^{\text{sep}}$ is some state which is separable with respect to the bipartition $AB|C$, $\varrho_{A|BC}^{\text{sep}}$ is separable with respect to $A|BC$ and $\varrho_{B|AC}^{\text{sep}}$ with respect to $B|AC$. A multipartite state that cannot be written in this form is called **genuinely multipartite entangled**.

Thus, a state like $\varrho_{AB} \otimes \varrho_C$ in which ϱ_{AB} is an entangled state, is entangled, but not genuinely multipartite entangled. In this thesis, we focus on genuine multipartite entanglement, as it is the strongest kind of entanglement, in which all particles are entangled with each other. For the sake of brevity, we will therefore simply use the term "entanglement" for multipartite states to refer to "genuine multipartite entanglement".

2.2 Detection of entanglement

Let us now consider the question how one can decide whether a state is entangled given its density matrix. It is not practicable to employ the definition of entanglement directly and to try to explicitly decompose the given state into separable states. If one is interested in showing that a state is entangled, one needs to prove that no such decomposition exists, which means that, in principle, one has to consider all possible decompositions. A proof that a state is separable can be done by giving a separable decomposition. However, in many cases this is not an easy task.

In this section, we will therefore present two ways to detect entanglement: the so-called PPT criterion in Sec. 2.2.1 and a class of operators, by the name of entanglement witnesses, which are commonly used in experiment to detect entanglement (Sec. 2.2.2). Then, we will specialize the latter notion to a certain subclass of entanglement witnesses. Note that Bell inequalities are another tool to detect entanglement which also plays a central role in this thesis. Since the underlying theory is a member of a whole class of so-called hidden-variable theories, Bell inequalities will be presented in an own section, namely Sections 2.4.1 and 2.4.2.

2.2.1 PPT criterion

First, we start by a definition that lies at the core of the separability criterion to be presented.

Definition 6. *Given a state*

$$\varrho = \sum_{i,j=0}^{d_A-1} \sum_{k,l=0}^{d_B-1} \varrho_{ij,kl} |i\rangle\langle j| \otimes |k\rangle\langle l| \qquad (2.8)$$

*of two particles, the first one of which is described by a Hilbert space of dimension d_A and d_B for the second particle. $|i\rangle \otimes |k\rangle$, $i = 1, \ldots, d_A - 1$, $k = 1, \ldots, d_B - 1$, is some product basis of the composite Hilbert space. Then, the **partial transpose** of ϱ with respect to subsystem A is defined as*

$$\varrho^{T_A} = \sum_{i,j=0}^{d_A-1} \sum_{k,l=0}^{d_B-1} \varrho_{ij,kl} |j\rangle\langle i| \otimes |k\rangle\langle l|. \qquad (2.9)$$

Note that the partial transpose ϱ^{T_A} depends on the product basis in which the partial transposition is performed. However, the spectrum of ϱ^{T_A} is basis-independent. Since the partial transpose can be calculated easily, the following necessary criterion for separability can be easily checked.

Theorem 7. *Any separable state ϱ_{sep} of two particles has a **positive partial transpose**, i.e. $\varrho_{\text{sep}}^{T_A}$ has no negative eigenvalues [30]. If the dimensions of the two Hilbert spaces are $d_A = 2$ and $d_B = 2$ or if they are $d_A = 2$ and $d_B = 3$, this positivity is also sufficient for separability.*

Thus, a negative eigenvalue of the partial transpose indicates that the respective state is entangled. Note that, for the sake of brevity, states with a positive partial transpose are usually said to be **PPT states**. Also, the fact that ϱ^{T_A} has no negative eigenvalues is denoted by $\varrho^{T_A} \geq 0$.

Proof. Per definition, any separable state can be written as in Eq. (2.5) and therefore its partial transpose with respect to A is

$$\varrho_{\text{sep}}^{T_A} = \sum_i p_i \left(\varrho_i^A\right)^T \otimes \varrho_i^B. \qquad (2.10)$$

Since the transposition does not affect the non-negativity of ϱ_i^A, $\left(\varrho_i^A\right)^T$ is also positive semidefinite and so is $\varrho_{\text{sep}}^{T_A}$. The proof of positive semidefiniteness being sufficient for separability for the given dimensions can be found in Ref. [31]. □

Clearly, the partial transposition can also be performed with respect to other subsystems. However, in the case of two particles, the partial transpose with respect to B does not yield new information, as

$$\varrho^{T_A} \geq 0 \Leftrightarrow \left(\varrho^{T_A}\right)^T = \varrho^{T_B} \geq 0. \qquad (2.11)$$

Here, it was used that an operator is positive if and only if its transpose is positive and that composition of partial transposition with respect to A and full transposition simply equals the partial transposition with respect to subsystem B.

2.2 Detection of entanglement

2.2.2 Entanglement witnesses

Another possibility to verify entanglement is the use of a special class of operators.

Definition 8. *A hermitian operator W is called an **entanglement witness** [31–33] if it fulfills the following two conditions:*

(i) $\text{Tr}(W \varrho_{\text{sep}}) \geq 0$ for all separable states ϱ_{sep}

(ii) $\text{Tr}(W \varrho_{\text{ent}}) < 0$ for at least one entangled state ϱ_{ent}

State ϱ_{ent} is then said to be detected by the entanglement witness W. Note that, while the last section covered the particle transpose for two-particle systems, the given definition of entanglement witnesses is independent from the number of particles. In the multipartite case, however, it is important to specify whether one refers to entanglement witnesses that are positive on fully separable states and detect entangled states or entanglement witnesses that are positive on biseparable states and detect genuinely multipartite entangled states. As mentioned before, we will focus on witnesses for genuine multipartite entanglement in this thesis.

An example for an entanglement witness that detects the singlet state

$$|\psi^-\rangle = \frac{1}{\sqrt{2}} (|01\rangle - |10\rangle) \tag{2.12}$$

is the witness

$$W_{\text{sing}} = \frac{1}{2}\mathbb{1} - |\psi^-\rangle\langle\psi^-| . \tag{2.13}$$

This operator is a witness, as it is positive on all PPT states and therefore positive on all separable states. This can be seen using the fact that the partial transposition obeys $\text{Tr}(W_{\text{sing}} \varrho^{T_A}) = \text{Tr}(W_{\text{sing}}^{T_A} \varrho)$. Moreover, $W_{\text{sing}}^{T_A}$ and ϱ are positive semidefinite operator. The partially transposed witness is positive semidefinite, as $\mathbb{1}$ does not change under partial transposition and a short calculation shows that $(|\psi^-\rangle\langle\psi^-|)^{T_A}$ has no eigenvalue larger than one half. This way to show that a given observable is a witness will often be applied in Sec. 5.6 in a similar way, but in a much more general setting.

Note that W_{sing} can also be in terms of Pauli matrices as

$$W_{\text{sing}} = \frac{1}{4} (X_1 X_2 + Y_1 Y_2 + Z_1 Z_2 + \mathbb{1}) . \tag{2.14}$$

Here, X_i denotes the Pauli matrix σ_x acting on the ith qubit. Analogously, Y_i and Z_i refer to the Pauli matrices σ_y and σ_z, respectively. Let us consider an experiment that aims at the preparation of a singlet state. Although the experimentally prepared state ϱ_{exp} will not exactly equal the singlet state due to experimental imperfections, ϱ_{exp} can be expected to be close to the singlet state. Therefore, the measured expectation value

2 Setting the stage

$$\langle W_{\text{sing}} \rangle = \frac{1}{4}(\langle X_1 X_2 \rangle + \langle Y_1 Y_2 \rangle + \langle Z_1 Z_2 \rangle + 1) \tag{2.15}$$

is likely to be negative and thus, the state ϱ_{exp} is proven to be entangled, independent from its actual exact form. Note that, here, $\langle A \rangle = \text{Tr}(A\varrho_{\text{exp}})$.

This illustrates a nice property of entanglement witnesses: While many separability criteria require knowledge of the whole density matrix, the witness' expectation value can usually be determined by measuring only a few expectation values, e.g., in the case of W_{sing} only three expectation values according to Eq. (2.15). On the contrary, determining a whole density matrix requires $4^n - 1$ measurements in the case of n qubits.

Note that, in many experiments such as experiments with photons, it is not possible to measure non-local observables like W_{sing}, but only local observables, which have a tensor product structure, such as $X_1 X_2$. This is why W_{sing} has to be decomposed into local observables before measuring it. Also, in experiments statistical errors have to be taken into account. Surprisingly, a larger number of expectation values in Eq. (2.15) does not necessarily imply a larger statistical error. This will be the subject of Sec. 3.

Moreover, entanglement witnesses can be interpreted in a geometrical way. Since $\text{Tr}(W\varrho)$ is linear in ϱ, the set of states for which $\text{Tr}(W\varrho) = 0$ defines a hyperplane that cuts the set of states in two parts. The first part consists of states with $\text{Tr}(W\varrho) < 0$. All of these states are detected by W and entangled. The second part contains states with $\text{Tr}(W\varrho) \geq 0$ and includes the set of separable states and some entangled states. This situation is shown in Fig. 2.1.

The figure also illustrates that there are always some states are not detected for a given witness W. On the other hand, there are "enough" witnesses to detect all entangled states [31].

Theorem 9. *For every entangled state ϱ_{ent} there is an entanglement witness that detects it.*

Proof. This property is easy to see as the set of separable states is convex and compact. Therefore, for any point that corresponds to an entangled state there is a hyperplane that separates this point from the set of separable states. Moreover, this hyperplane defines an entanglement witness. □

Moreover, Fig. 2.1 also shows what is meant when one witness is said to be finer than another one.

Definition 10. *Given two entanglement witnesses W_1 and W_2. If the set of states detect by W_1 is a strict subset of the states detected by W_2, then W_2 is said to be **finer** than W_1.*

Lemma 11. *If entanglement witness W_2 is finer than W_1, there is a positive hermitian operator P, such that $W_1 = W_2 + P$ [33].*

Naturally, this notion gives rise to the idea of "finest" witnesses.

Definition 12. *An entanglement witness W is called **optimal**, if there is no witness that is finer than W.*

2.2 Detection of entanglement

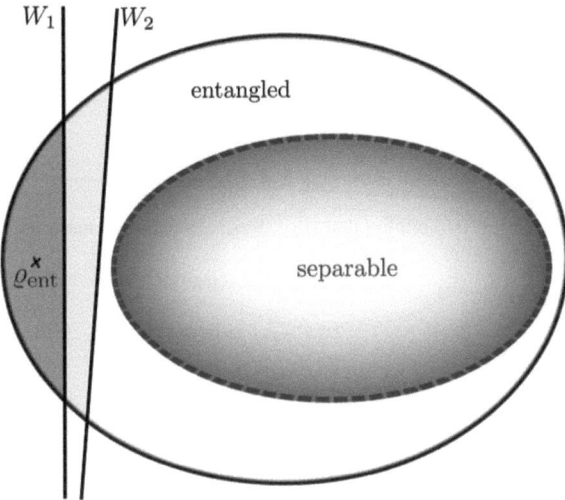

Figure 2.1: Entanglement witnesses define a hyperplane in the set of states. Here, two witnesses, W_1 and W_2, are visualized by a line. The set of states that each of them detects is shaded in gray. Both of them detect ϱ_{ent}, but W_2 is finer than W_1. The set of all states is visualized by an ellipse with black border, while the set of separable states is shown in blue. Both sets are convex. Any state outside of the blue set of separable states is entangled by definition.

Note that, the optimal entanglement witness for a given state ϱ is not unique. Also, the fact that there is a separable state ϱ_{sep} with $\text{Tr}(W\varrho_{\text{sep}})$, i.e. that W touches the set of separable states, does not imply it is an optimal witness in the sense defined above. In Ref. [33], it has been shown how to optimize witnesses.

Finally, note that witness operators can also be defined for other properties besides entanglement. Instead of witnessing that a state is not separable, an operator that is positive some other convex set of states could witness that a given state does not belong to this convex set. Since the set of PPT states is also convex, an example are operators that detect whether a state is not PPT. This will be one of the starting points of Sec. 4.

Decomposable witnesses

Let us now return to the case of two particles. In general, it is not easy to verify that a given operator is an entanglement witness. However, there is a certain class of witnesses that have a structure which immediately allows one to see that they are positive on all separable states.

Definition 13. *In a bipartite system, a witness W is called **decomposable**, if there exists two positive hermitian operators P and Q such that*

$$W = P + Q^{T_A}, \qquad (2.16)$$

where T_A is the partial transposition with respect to subsystem A [33].

It is easy to see that Eq. (2.16) implies that W is positive on all PPT states. Let ϱ_{ppt} be a PPT state. Then, for a decomposable witness W,

$$\begin{aligned}\text{Tr}(W\varrho_{\text{ppt}}) &= \text{Tr}(P\varrho_{\text{ppt}}) + \text{Tr}(Q^{T_A}\varrho_{\text{ppt}}) \\ &= \text{Tr}(P\varrho_{\text{ppt}}) + \text{Tr}(Q\varrho_{\text{ppt}}^{T_A}) \geq 0\,.\end{aligned} \qquad (2.17)$$

Here, $\text{Tr}(AB^{T_A}) = \text{Tr}(A^{T_A}B)$ has been used in the second line. The positivity follows from the fact that P, Q and ϱ_{ppt} are positive operators and ϱ_{ppt} remains positive under partial transposition. According to Theorem 7, the set of separable states is a subset of the PPT states and therefore W is positive on all separable states. Note that the witness of Eq. (2.13) is also a decomposable witness with $P = 0$.

2.3 Quantification of entanglement

The question whether a state is entangled naturally leads to the question of how much it is entangled. The quantity of entanglement contained in a state is usually measured by so-called entanglement measures or entanglement monotones. While there are several entanglement monotones for bipartite systems, entanglement monotones for multipartite systems are much less numerous. In Sec. 2.3.1, we will give an overview over the properties an entanglement monotone has to fulfill in the bipartite case and briefly mention some examples. Then, we will present two entanglement measures for genuine multipartite entanglement in Sec. 2.3.2.

2.3.1 Bipartite entanglement measures

In order to define what an entanglement monotone is, we need to the notion of **local operations and classical communication** (LOCC). An LOCC operation on some state ϱ is an operation in which Alice and Bob can perform any kind of local operation and communicate over a classical channel. These local operations include measurements, unitary operations, attaching an ancilla system and performing operations on the ancilla system or the combined system of ancilla and original particle. Alice and Bob can communicate before and after any such operations and, in this way, perform operations that depend on the outcome of the previous operation of either of the two.

2.3 Quantification of entanglement

Due to this complexity of possible operations and protocols, there is no simple mathematical description of LOCC known. However, when one needs to optimize over all possible LOCC operations, one often optimizes over a set of operations that is easier to characterize and also contains the set of LOCC operations [34]. These are the so-called **separable operations**. The set of separable operations are all operations Λ_{sep} that map a given state ϱ onto

$$\Lambda_{\text{sep}}(\varrho) = \sum_i p_i \frac{K_i \varrho K_i^\dagger}{\text{Tr}(K_i \varrho K_i^\dagger)}, \qquad (2.18)$$

where $K_i = A_i \otimes B_i$, $\sum_i K_i^\dagger K_i = \mathbb{1}$ and $p_i = \text{Tr}(K_i \varrho K_i^\dagger)$ is the probability with which the result of the operation is $K_i \varrho K_i^\dagger$.

Since LOCC operations are a subset of the separable operations, an maximization over separable operations provides an upper bound for maximizations over all LOCC operations. Note that, also for LOCC operations, the outcome is, in general, not deterministic. Therefore, the possible results of the operation are given by some states ϱ_i that occur according to some probabilities p_i.

An entanglement monotone is given by a function

$$\begin{aligned} E : \mathcal{D}(\mathcal{H}) &\to \mathbb{R}_0^+ \\ \varrho &\mapsto E(\varrho), \end{aligned} \qquad (2.19)$$

where $\mathcal{D}(\mathcal{H})$ is the set of density operators and $\mathbb{R}_0^+$ is the set of non-negative real numbers, which additionally has to fulfill certain conditions. However, there are differences in the literature as to what conditions these are. Moreover, although an entanglement measure usually fulfills more conditions than a monotone, these two terms are often used interchangeably. For these reasons, we first provide a list of conditions that are used in the literature [35, 36].

(i) $E(\varrho_{\text{sep}}) = 0$ for all separable states ϱ_{sep}.

(ii) Applying LOCC operations to any state ϱ does not increase the value of E, i.e.

$$E(\Lambda_{\text{LOCC}}(\varrho)) \leq E(\varrho) \text{ for all } \varrho. \qquad (2.20)$$

(iii) E is convex, since mixing of two states should not increase the amount of entanglement,

$$E\left(\sum_i p_i \varrho_i\right) \leq \sum_i p_i E(\varrho_i). \qquad (2.21)$$

While one always requires an entanglement monotone to fulfill (i), condition (ii) is often replaced by

(ii') Applying LOCC operations to any state ϱ does not increase the value of E in average, i.e.

$$\sum_i p_i E(\varrho_i) \leq E(\varrho) \text{ for all } \varrho. \qquad (2.22)$$

Note that (ii') implies condition (ii). Moreover, entanglement monotones should not change under local basis changes. However, since local basis changes are included in the set of LOCC operations and invertible, the invariance under local basis changes follows from (ii). Also, condition (iii) is sometimes not postulated for an entanglement monotone.

In contrast, an entanglement measure is usually required to fulfill the following condition in addition to the ones presented above.

(iv) For any pure state $|\psi\rangle$,
$$E(|\psi\rangle\langle\psi|) = -\text{Tr}[\varrho_A \log_2(\varrho_A)], \qquad (2.23)$$
where $\varrho_A = \text{Tr}_A(|\psi\rangle\langle\psi|)$.

Examples

There are many entanglement measures which have an operational meaning, such as the entanglement cost $E_C(\varrho)$, which is basically given by the minimal number of singlet states needed to create a large number of ϱ via LOCC. The entanglement of distillation $E(\varrho)$ is given by the number of singlets that one can obtain from a large number of copies of ϱ [37]. The entanglement of formation $E_F(\varrho)$ extends the quantity of Eq. (2.23) to mixed states via the so-called convex-roof construction
$$E_F(\varrho) = \inf_{p_i, |\phi_i\rangle} \sum_i p_i S[\text{Tr}_A(|\phi_i\rangle\langle\phi_i|)]. \qquad (2.24)$$

Here, the infimum is taken over all possible decompositions $\varrho = \sum_i p_i |\phi_i\rangle\langle\phi_i|$.

However, all of the mentioned entanglement measures are, in general, difficult to compute. An easily computable measure is given by the violation of the PPT criterion (cf. Theorem 7).

Definition 14. *For any bipartite state ϱ, its **negativity** [38] is defined as*
$$N(\varrho) = \frac{\|\varrho^{T_A}\|_1 - 1}{2}. \qquad (2.25)$$

Using that $\|\varrho^{T_A}\|_1$ is the sum over all absolute values of the eigenvalues of ϱ^{T_A} and the normalization of ϱ, one sees that the negativity can be calculated by summing up the absolute values of ϱ^{T_A}'s negative eigenvalues. The negativity is also convex.

For completeness, we briefly mention the **concurrence** [39, 40], which, for pure states, is defined as
$$C(|\psi\rangle) = \sqrt{2[1 - \text{Tr}(\varrho_A^2)]}, \text{ where } \varrho_A = \text{Tr}_A(|\psi\rangle\langle\psi|). \qquad (2.26)$$

Also, its convex-roof extension [cf. Eq. (2.24)] to mixed states of two qubits can be calculated analytically [41].

2.3 Quantification of entanglement

2.3.2 Multipartite entanglement measures

In the multipartite case, one has to distinguish between measures that vanish on all fully separable states and measures that vanish on all biseparable states. While the first detect multipartite entanglement (as opposed to full separability), the latter detect genuine multipartite entanglement. It is also worth mentioning that, as there is no general consensus on the axioms that an entanglement monotone has to fulfill in the bipartite case, there is also no general agreement in the multipartite case. However, a natural set of operations under which such a quantity should not increase are operations that are fully local, i.e. a full tensor product of all parties, and classical communication between all of them.

The **geometric measure of entanglement** [42–44] E_G of a pure state $|\psi\rangle$ is given by its maximal overlap with a fully separable state,

$$E_G(|\psi\rangle) = 1 - \sup_{|\phi\rangle=|a\rangle|b\rangle|c\rangle\ldots} |\langle\psi|\phi\rangle|^2 . \tag{2.27}$$

It can be extended to mixed states using the convex-roof construction. Although this measure only vanishes on fully separable states, the idea of such distance measures can also be used to construct measures for genuine multipartite entanglement. For example, the **relative entropy of entanglement** [45] of a state ϱ is defined by

$$E_R(\varrho) = \inf_\sigma S(\varrho \parallel \sigma) , \tag{2.28}$$

where $S(\varrho \parallel \sigma) = \text{Tr}[\varrho\log(\varrho) - \varrho\log(\sigma)]$ is the so-called relative entropy.

Similarly, one can define the robustness $R(\varrho)$ of a state as

$$R(\varrho) = \min\{s \ : \ \frac{\varrho + s\sigma}{1+s}\} . \tag{2.29}$$

Here, σ is an arbitrary separable state [46]. It is also possible to consider the quantity of Eq. (2.29) for σ being an arbitrary state or identity [47].

For three qubits, there is the so-called three-tangle τ [48] which is for pure states defined by

$$\tau(|\psi\rangle) = C^2_{A|BC}(|\psi\rangle) - C^2_{AB}(\varrho_{AB}) - C^2_{AC}(\varrho_A C) \tag{2.30}$$

and for mixed states through the convex-roof construction. Here, $C_{A|BC}$ is the concurrence [cf. Eq. (2.26)] between system A and the combined system BC, $C_{AB}(\varrho_{AB})$ is the concurrence between A and B and analogously for $C_{AC}(\varrho_{AC})$. Note that the three-tangle vanishes not only on all biseparable states, but also on all states that can be created from the state

$$|W\rangle = \frac{1}{\sqrt{3}}(|001\rangle + |010\rangle + |100\rangle) \tag{2.31}$$

by stochastic LOCC operations [20]. In stochastic LOCC operations, one performs LOCC operations and chooses, out of the many possible resulting states that occur with a non-zero probability, one of these possible states. The three-tangle can be calculated analytically for important cases [49] and can be generalized to a higher number of particles [50, 51].

For general mixed states, the aforementioned measures are difficult to compute. In Sec. 4, a generalization of the negativity to the multipartite case, which can easily be computed numerically for an arbitrary state, will be presented.

2.4 Hidden-variable theories

Even before the field of quantum information emerged in the late 1980s, entanglement played in important role in understanding nature's properties. Einstein, Podolski and Rosen suggested that the probabilistic character of quantum mechanics only stems from the fact that the theory is not complete and that there is a complete theory behind it which predicts the outcomes of all possible measurements [1]. Without specifying the exact form of such a theory, Bell subsequently introduced a framework in which such theories can be described, namely what we now call **hidden-variable theories**.

Let us first assume two particles controlled by Alice and Bob, respectively. In the framework of hidden variables, the probability that Alice obtains the outcome α and Bob the outcome β when Alice is measuring observable A and Bob is measuring observables B given that the hidden variable is λ is denoted by $p_\lambda(\alpha, \beta|A, B)$. Note that the exact form of the hidden variable is not specified here. It can, in principle, be anything from a scalar to a high-dimensional vector. The expectation value of the observable AB given that the hidden variable is λ is then calculated according to

$$\langle AB \rangle_\lambda = \sum_{\alpha, b} \alpha\beta \, p_\lambda(\alpha, \beta|A, B) \,. \tag{2.32}$$

As, in reality, we lack knowledge of the hidden variable, we actually measure the expectation value

$$\langle AB \rangle = \int d\lambda \varrho(\lambda) \sum_{\alpha, \beta} \alpha\beta \, p_\lambda(\alpha, \beta|A, B) \tag{2.33}$$

in experiment. Here, $\varrho(\lambda)$ is a probability density on the hidden variable λ.

2.4.1 Bipartite Bell inequalities

Besides the existence of hidden variables, Bell theories also assume locality. In other words, one considers hidden-variable theories in which the expectation values of Eq. (2.33) can always be written in terms of probability distributions that factorize, i.e. which obey

$$p_\lambda(\alpha, \beta|A, B) = p_\lambda(\alpha|A) p_\lambda(\beta|B) \tag{2.34}$$

for any two observables A_1 and B_1 and fixed λ. Hidden-variable theories that obey Eq. (2.34) are called **local hidden-variable theories** (LHV theories). Note that locality implies that Alice's choice

2.4 Hidden-variable theories

of observable cannot affect Bob's outcome probabilities (no-signalling), i.e., that for any two observables B and B' that Bob measures (and any observable A and outcome a of Alice), the probabilities obey

$$p_\lambda(\alpha|A, B) = p_\lambda(\alpha|A, B'), \qquad (2.35)$$

where $p_\lambda(\alpha|A, B) = \sum_b p_\lambda(\alpha, \beta|A, B)$.

The most commonly used Bell inequality for two particles is the Clauser-Horne-Shimony-Holt (CHSH) inequality [52]. It states that in any LHV theory, the inequality

$$\langle AB \rangle + \langle AB' \rangle + \langle A'B \rangle - \langle A'B' \rangle \leq 2 \qquad (2.36)$$

holds for any observables A, A', B and B'. In quantum mechanics, this inequality can be violated by choosing $A = -X$, $A' = -Y$, $B = (X - Y)/\sqrt{2}$, $B' = (X + Y)/\sqrt{2}$ and the singlet state of Eq. (2.12). For these choices, the left-hand side of Eq. (2.36) equals $2\sqrt{2}$.

Bell inequalities can also be used for entanglement detection, as any state that violates a Bell inequality must be entangled. This can be seen by noting that for any separable state $\varrho_{\text{sep}} = \sum_i p_i \varrho_i^A \otimes \varrho_i^B$, one has

$$\langle AB \rangle = \sum_i p_i \, \text{Tr}(A\varrho_i^A) \text{Tr}(B\varrho_i^B) \,. \qquad (2.37)$$

Then, however, ϱ_i^A and ϱ_i^B can be interpreted as different values of the hidden variable. This defines an LHV model, where the locality can be seen by Eq. (2.37), and therefore, it cannot violate a Bell inequality. However, the converse is not true: There are some entangled states that do not violate any Bell inequality [4, 53]. Finally, it is worth mentioning that Bell inequalities are, in contrast to entanglement witnesses, tools to detect entanglement independent from the observables actually measured in experiment. As experimental implementations are never perfect, one might measure observables different from the ones that one intended to measure. However, since Eq. (2.36) holds for any observables, a violation of this inequality would still indicate the presence of entanglement.

2.4.2 Multipartite Bell inequalities

Analogously to multipartite entanglement, there are two different notions of non-locality in the multipartite case. Let us first consider probability distributions which factorize fully and have the form

$$p_\lambda(\alpha, \beta, \gamma|A, B, C) = p_\lambda(\alpha|A)p_\lambda(\beta|B)p_\lambda(\gamma|C) \,. \qquad (2.38)$$

Probability distributions of this kind, which one might call "fully local", obey the **Mermin inequality** [54]. For three qubits, the Mermin inequality is given by

$$\langle ABC \rangle - \langle AB'C' \rangle - \langle A'BC' \rangle - \langle A'B'C \rangle \leq 2 \,, \qquad (2.39)$$

15

where A, A', B, B', C and C' are arbitrary observables. Note that, from now on, we will present Bell inequalities with the quantum mechanical observables that yield the largest violation already plugged in, since this allows for a more compact notation. In the case of the Mermin inequality, we then have

$$\langle X_1 X_2 X_3 \rangle - \langle X_1 Y_2 Y_3 \rangle - \langle Y_1 X_2 Y_3 \rangle - \langle Y_1 Y_2 X_3 \rangle \leq 2 \,. \tag{2.40}$$

This inequality is maximally violated for the three-qubit Greenberger-Horne-Zeilinger (GHZ) state

$$|GHZ_3\rangle = \frac{1}{\sqrt{2}}(|000\rangle + |111\rangle)\,. \tag{2.41}$$

In this case, the left-hand side has a value of 4, since every term has an absolute value of 1 with the appropriate sign. This observation was also the basis for the argument by Greenberger, Horne and Zeilinger [55] who argue that the GHZ state contradicts realism in the sense of Einstein, Podolsky and Rosen [1].

For n qubits, the Mermin inequality is given by

$$\langle X_1 X_2 X_3 X_4 X_5 \ldots X_n \rangle - \sum_{\text{perms}} \langle Y_1 Y_2 X_3 X_4 X_5 \ldots X_n \rangle$$
$$+ \sum_{\text{perms}} \langle Y_1 Y_2 Y_3 Y_4 X_5 \ldots X_n \rangle - \cdots \leq \begin{cases} 2^{n/2}, \text{ for even } n, \\ 2^{(n-1)/2}, \text{ for odd } n \,. \end{cases} \tag{2.42}$$

Here, $\sum_{\text{perms}}$ indicates a sum over all permutations of all qubits that lead to distinct terms. Again, the Mermin inequality holds also for an arbitrary choice of observables. The maximal violation is obtained for the n-qubit GHZ state

$$|GHZ_n\rangle = \frac{1}{\sqrt{2}}(|0\ldots0\rangle + |1\ldots1\rangle)\,, \tag{2.43}$$

for which the left-hand side reaches a value of 2^{n-1}.

Another example that holds for the same kind of non-locality is the **Ardehali inequality** [56], which is given by

2.4 Hidden-variable theories

$$\begin{aligned}
&\Big[\langle A_1 X_2 X_3 X_4 X_5 X_6 X_7 \ldots X_n\rangle + \langle B_1 X_2 X_3 X_4 X_5 X_6 X_7 \ldots X_n\rangle \\
&- \sum_{\text{perms}(2,\ldots,n)} (\langle A_1 Y_2 X_3 X_4 X_5 X_6 X_7 \ldots X_n\rangle - \langle B_1 Y_2 X_3 X_4 X_5 X_6 X_7 \ldots X_n\rangle) \\
&- \sum_{\text{perms}(2,\ldots,n)} (\langle A_1 Y_2 Y_3 X_4 X_5 X_6 X_7 \ldots X_n\rangle + \langle B_1 Y_2 Y_3 X_4 X_5 X_6 X_7 \ldots X_n\rangle) \\
&+ \sum_{\text{perms}(2,\ldots,n)} (\langle A_1 Y_2 Y_3 Y_4 X_5 X_6 X_7 \ldots X_n\rangle - \langle B_1 Y_2 Y_3 Y_4 X_5 X_6 X_7 \ldots X_n\rangle) \\
&+ \sum_{\text{perms}(2,\ldots,n)} (\langle A_1 Y_2 Y_3 Y_4 Y_5 X_6 X_7 \ldots X_n\rangle + \langle B_1 Y_2 Y_3 Y_4 Y_5 X_6 X_7 \ldots X_n\rangle) \\
&- \sum_{\text{perms}(2,\ldots,n)} (\langle A_1 Y_2 Y_3 Y_4 Y_5 Y_6 X_7 \ldots X_n\rangle - \langle B_1 Y_2 Y_3 Y_4 Y_5 Y_6 X_7 \ldots X_n\rangle) - \ldots \Big]/\sqrt{2} \\
&\leq \begin{cases} 2^{(n-1)/2}, & \text{for even } n, \\ 2^{n/2}, & \text{for odd } n. \end{cases}
\end{aligned} \qquad (2.44)$$

Here, $\sum_{\text{perms}(2,\ldots,n)}$ denotes a sum over all permutations of qubits 2 to n that yield distinct observables. Moreover, $A = (X_1 + Y_1)/\sqrt{2}$ and $A' = (X_1 - Y_1)/\sqrt{2}$. Again, the Ardehali inequality holds for arbitrary observables, but with the above observables and the GHZ state, the quantum mechanical violation is maximal and equals 2^{n-1}. Note that the quantum mechanical operators on the left-hand side of Eqs (2.42) and (2.44) are the same, but the bound on LHV models, i.e., the right-hand side differs.

As for entanglement, there is also a notion of **genuine multipartite non-locality**. For three qubits, any probability distribution that cannot be written as

$$\begin{aligned}
&p_\lambda(\alpha, \beta, \gamma | A, B, C, \ldots) \\
&= q_1 p_\lambda(\alpha | A) p_\lambda(\beta, \gamma | B, C) + q_2 p_\lambda(\beta | B) p_\lambda(\alpha, \gamma | A, C) + q_3 p_\lambda(\gamma | C) p_\lambda(\alpha, \beta | A, B),
\end{aligned} \qquad (2.45)$$

where $\sum_i q_i = 1$ and $q_i \geq 0$, is called genuine multipartite non-local. Any probability distribution that *is* of the form of Eq. (2.45) obeys the Svetlichny inequality [57]

$$\langle ABC\rangle + \langle AB'C\rangle + \langle ABC'\rangle - \langle AB'C'\rangle + \langle A'BC\rangle - \langle A'B'C\rangle - \langle A'BC'\rangle - \langle A'B'C'\rangle \leq 4. \qquad (2.46)$$

The quantum mechanical violation is maximal for the GHZ state [cf. Eq. (2.41)] of three qubits and equals $4\sqrt{2}$ for the choice $A = -X$, $A' = Y$, $B = (X+Y)/\sqrt{2}$, $B' = (X-Y)/\sqrt{2}$, $C = -X$ and $C' = Y$ [58]. One way to prove that Eq. (2.46) holds for genuinely non-local models is based on the realization that the inequality is a sum of two CHSH inequalities. For example, all expectation values containing A form a CHSH inequality on parties two and three and the same for all terms that include

A'. Moreover, since Eq. (2.46) is invariant under any permutation of particles, it has this form on any two qubits.

Also, Svetlichny's inequality has been generalized to an arbitrary number of qubits [15].

2.4.3 Leggett inequalities

In 2003, Leggett introduced another class of hidden-variable theories, replacing the assumption of locality in Bell inequalities by an assumption on the marginal probability distributions [28, 59]. More precisely, consider, for a fixed hidden variable λ, the probability distributions $p_\lambda(\alpha, \beta|A, B)$ on the measurement outcomes α and β given that A and B have been measured, and their marginal

$$p_\lambda(\alpha|A) = \sum_b p_\lambda(\alpha, \beta|A, B). \tag{2.47}$$

Here, by writing $p_\lambda(\alpha|A)$ instead of $p_\lambda(\alpha|A, B)$, we have also assumed no-signalling [cf. Eq. (2.35)]. Now, consider the set $\mathcal{L}$ of all probability distributions whose marginals obey the following condition: For every observable A, there exists a unit vector $\vec{a} \in \mathbb{R}^3$, such that

$$\langle A \rangle_\lambda = \sum_\alpha \alpha p_\lambda(\alpha|A) = \vec{\lambda}\vec{a}. \tag{2.48}$$

Here, the unit vector $\vec{\lambda} \in \mathbb{R}^3$ corresponds to the hidden variable and is fixed. Note that the first equality sign holds due to the definition of an expectation value and the second equality is the actual assumption that we impose. Then, a **Leggett model** is a hidden variable model in which all probability distributions are convex combinations of distributions whose marginals obey Eq. (2.48), i.e. all probability distributions are in the convex hull of $\mathcal{L}$.

Equation (2.48) can be interpreted as follows: Every one-qubit quantum state can be written as $\varrho = \frac{1}{2}\left(1 + \sum_{i=1}^{3} \lambda_i \sigma_i\right)$, where σ_i are the three Pauli matrices and $\lambda_i \in \mathbb{R}^3$ is of unit length for pure states and of length less than one for mixed states. As we can also write the observable $A = \sum_{i=1}^{3} a_i \sigma_i$ in the Pauli basis using a three-dimensional real vector $\vec{a}$, we have $\text{Tr}(A\varrho) = \vec{\lambda}\vec{a}$ with $|\vec{\lambda}| = 1$ for pure states. Thus, Eq. (2.48) says that every reduced one-qubit state (of Alice) behaves like a pure state (for a given λ). Naturally, one can postulate the same condition on Bob's side. However, in the following, we only need to make use of such a condition on Alice's side.

Following the derivation in Ref. [60], we consider the probability $p_\lambda(\alpha_1, \beta_1|A, B)$ of measuring outcomes α_1 and β_1, when the observables A and B are measured, for a fixed hidden variable λ. This probability is given by

$$p_\lambda(\alpha_1, \beta_1|A, B) = \frac{1}{4}(1 + \alpha_1 \langle A \rangle + \beta_1 \langle B \rangle + \alpha_1 \beta_1 \langle AB \rangle). \tag{2.49}$$

Here, $\langle A \rangle$ is the expectation value of A on Alice's side when, at the same time, Bob is measuring B on his side (and vice versa for $\langle B \rangle$). However, assuming no-signalling [cf. Eq. (2.35)] implies that

2.4 Hidden-variable theories

$\langle A \rangle$ is independent from Bob's choice of observables and justifies the notation $\langle A \rangle$. If we invoke this assumption and consider a second probability distribution of measuring outcomes α_2 and β_2 for the observables A and B', then the sum of these distributions is given by

$$p_\lambda(\alpha_1, \beta_1 | A, B) + p_\lambda(\alpha_2, \beta_2 | A, B')$$
$$= \frac{1}{4}[2 + (\alpha_1 + \alpha_2)\langle A \rangle + \beta_1 \langle B \rangle + \beta_2 \langle B' \rangle + \alpha_1 \beta_1 \langle AB \rangle + \alpha_2 \beta_2 \langle AB' \rangle] \,. \tag{2.50}$$

Positivity of these probability distributions for any value of $\alpha_1, \alpha_2, \beta_1, \beta_2 \in \{-1, +1\}$ implies

$$|\langle AB \rangle_\lambda + \langle AB' \rangle_\lambda| \leq 2 - |\langle B \rangle_\lambda + \langle B' \rangle_\lambda| \,. \tag{2.51}$$

Integration $\int d\lambda \, \varrho(\lambda)$, where $\varrho(\lambda)$ is some probability density function on the hidden variables, and use of $|\int .| \leq \int |.|$ yields

$$|\langle AB \rangle + \langle AB' \rangle| \leq 2 - \int d\lambda \, \varrho(\lambda) |\langle B \rangle_\lambda - \langle B' \rangle_\lambda| \,. \tag{2.52}$$

Note that the left-hand side is now independent from λ, as we have used the definition of Eq. (2.33). Finally, we plug in the assumption of Eq. (2.48), which results in

$$|\langle AB \rangle + \langle AB' \rangle| \leq 2 - \int d\lambda \, \varrho(\lambda) |\vec{\lambda}\vec{b} - \vec{\lambda}\vec{b'}| \,, \tag{2.53}$$

where $\vec{b}, \vec{b'} \in \mathbb{R}^3$ are unit vectors. This equation contains three observables, namely A, B and B'. Consider an experiment in which such a triple of observables is measured for three different settings, resulting in the observables A_i, B_i, B'_i, $i = 1, 2, 3$. As Bob cannot measure B_i and B'_i at the same time, this actually requires six different runs in experiment — three for A_i, B_i and three for A_i, B'_i. To A_i, we associate the unit vector $\vec{a}_i$ and to B_i and B'_i, we associate $\vec{b}_i$ and $\vec{b'}_i$, respectively. Now, we define the unit vector $\vec{e}_i$ by $\vec{b}_i - \vec{b'}_i = 2\sin(\varphi/2)\vec{e}_i$, where φ_i is the angle between $\vec{b}_i$ and $\vec{b'}_i$. We consider the case in which $\varphi = \varphi_1 = \varphi_2 = \varphi_3$ and sum Eq. (2.53) over all three settings. This results in

$$\frac{1}{3}\sum_{i=1}^{3} |\langle A_i B_i \rangle + \langle A_i B'_i \rangle| \leq 2 - \frac{2}{3}|\sin(\varphi/2)| \int d\lambda \, \varrho(\lambda) \sum_{i=1}^{3} |\vec{\lambda}\vec{e}_i| \,. \tag{2.54}$$

If $\{\vec{e}_1, \vec{e}_2, \vec{e}_3\}$ are orthonormal, then, for any $\vec{\lambda}$, we have $|\vec{\lambda}\vec{e}_i| \geq 1$ and therefore

$$\frac{1}{3}\sum_{i=1}^{3} |\langle A_i B_i \rangle + \langle A_i B'_i \rangle| \leq 2 - \frac{2}{3}|\sin(\varphi/2)| \,, \tag{2.55}$$

which holds for all Leggett models. Quantum mechanics, however, violates this inequality. For the

19

singlet state, we have $\langle A_i B_i \rangle = -\vec{a_i}\vec{b_i}$ and therefore

$$2|\cos(\varphi/2)| \leq 2 - \frac{2}{3}|\sin(\varphi/2)|, \quad (2.56)$$

which is maximally violated for $\varphi \approx 0.680$, where the left-hand side of Eq. (2.56) is approximately 1.06 times as big as the right-hand side.

One can therefore conclude that quantum mechanics cannot be characterized by a hidden-variable model in which the single qubits behave as if they were in a pure state. In Sec. 7, we will consider the question of how one can define Leggett models and construct the according inequalities in the multipartite case.

2.5 Graph states

In this section, we will introduce graph states, a special class of states that can be described by an efficient and useful formalism. Graph states play an important role for tasks like measurement-based quantum computation [7, 8] or quantum error correction [61–63]. These states have several interesting properties, for instance they are relatively robust against decoherence and violate certain Bell inequalities maximally [64]. Recently, several experiments succeeded in preparing graph states of several qubits with photons [25, 65–71], and also the theory of entanglement detection for such experiments has been investigated in a number of papers [72, 73].

Examples for graph states are the 2D cluster state [74] used for universal measurement-based quantum computation, the GHZ state of Eq. (2.43) [55], the ring cluster and the linear cluster state. An introduction to graph states can be found in Ref. [64]. Finally, note that a considerable part of this section is taken from Ref. [75].

In this introductory section, we will start with the definition in Sec. 2.5.1, then define the so-called graph state basis (cf. Sec. 2.5.2) and finally consider the application of local unitary operations to graph states in Sec. 2.5.3.

2.5.1 Definition

Graph states are defined by mathematical graphs in the following way. Given a graph $G = (V, E)$ that is defined by a set V of vertices which correspond to qubits and a set E of edges that connect some of these vertices (cf. the examples in Fig. 2.2). We denote the number of vertices by n. Then, one can define a set of n operators

$$g_i = X_i \prod_{k \in \mathcal{N}(i)} Z_k, \quad i = 1, \ldots, n, \quad (2.57)$$

where $\mathcal{N}(i)$ is the **neighborhood** of qubit i, i.e., the set of all qubits that are connected to qubit i by an edge.

2.5 Graph states

The operators g_i commute and generate a set $\mathcal{S}$ of so-called **stabilizer operators** which consists of 2^n elements, i.e.,

$$\mathcal{S} = \{S_1, \ldots, S_{2^n}\} = \left\{\prod_{i=1}^{n} g_i^{x_i} \Big| \vec{x} \in \{0,1\}^n \right\}. \tag{2.58}$$

Here, $\vec{x}$ is a vector of length n containing only zeros and ones. This means that every operator $S_i \in \mathcal{S}$ can be written as a product of some generators g_i, in which every generators appears once or not at all. Note that the identity operator $\mathbb{1}$ is also contained in $\mathcal{S}$, as it is obtained for $\vec{x}$ being a vector of only zeros. Since $g_i g_i = \mathbb{1}$, $i = 1, \ldots, n$, the product of g_i with itself is included in the definition of Eq. (2.58) and $\mathcal{S}$ is closed under multiplication. Now, we can define the corresponding graph state $|G\rangle$.

Definition 15. *Given a graph $G = (V, E)$. The corresponding **graph state** $|G\rangle$ is uniquely defined by*

$$g_i |G\rangle = |G\rangle, \ \forall\, i = 1, \ldots, n, \tag{2.59}$$

where the g_i are defined by Eq. (2.57).

Thus, $|G\rangle$ is the common eigenstate of all generators g_i with eigenvalue $+1$. As an example, let us consider the four Bell states

$$|\phi^+\rangle = \frac{1}{\sqrt{2}}(|00\rangle + |11\rangle), \tag{2.60}$$

$$|\phi^-\rangle = \frac{1}{\sqrt{2}}(|00\rangle - |11\rangle), \tag{2.61}$$

$$|\psi^+\rangle = \frac{1}{\sqrt{2}}(|01\rangle + |10\rangle), \tag{2.62}$$

$$|\psi^-\rangle = \frac{1}{\sqrt{2}}(|01\rangle - |10\rangle). \tag{2.63}$$

Here, the last state is the singlet state that has already been introduced in Eq. (2.12). The associated graph is shown in Fig. 2.2 as No. 1. According to Eq. (2.57), the generators of the state's stabilizer group are

$$g_1 = X_1 Z_2, \ g_2 = Z_1 X_2. \tag{2.64}$$

Then, the state

$$|G_1\rangle = \frac{1}{\sqrt{2}}(|+0\rangle + |-1\rangle) \tag{2.65}$$

is an eigenstate of g_1 and g_2 with eigenvalue $+1$. Here, $|\pm\rangle = (|0\rangle \pm |1\rangle)/\sqrt{2}$ are the eigenvectors of X for eigenvalue ± 1 and $|0\rangle$ and $|1\rangle$ are the eigenvectors of Z. It is therefore easy to check that $|G_1\rangle$ is really an eigenvector for eigenvalue $+1$ for the two g_i, if one keeps in mind that X acts as a flip operator on $|0\rangle$ and $|1\rangle$ and Z as a flip operator on $|\pm\rangle$.

One notes that $|G_1\rangle$ does not equal any Bell state exactly. However, by applying appropriate local unitary operations, one can transform the state of Eq. (2.65) to any of the four Bell states. This is why the graph No. 1 in Fig. 2.2 is said to belong to a Bell state.

Another example is the GHZ state of three qubits [cf. Eq. (2.41)]. Its associated graph is shown in Fig. 2.2 as No. 2. Therefore, the generators of the stabilizer group are given by

$$g_1 = X_1 Z_2, \; g_2 = Z_1 X_2 Z_3, \; g_3 = Z_2 X_3 \,. \tag{2.66}$$

Then, the state

$$|G_2\rangle = \frac{1}{\sqrt{2}}(|-1-\rangle + |+0+\rangle) \tag{2.67}$$

is an eigenstate of all of these generators. Also here, one can see that the three-qubit GHZ state in its standard form as in Eq. (2.41) is different from the state $|G_2\rangle$ that is associated to graph No. 2 of Fig. 2.2. However, after a local basis change, $|G_2\rangle$ and the three-qubit GHZ state coincide.

2.5.2 Graph state basis

Every graph also defines a basis of orthonormal states.

Definition 16. *Given a graph a* $G = (V, E)$ *and the corresponding generators* g_i *[cf. Eq. (2.57)]. The* **graph state basis** *of this graph is the set of states* $\{|a_1 \ldots a_n\rangle_G | a_i \in \{0,1\}\}$ *which obey*

$$g_i |a_1 \ldots a_n\rangle_G = (-1)^{a_i} |a_1 \ldots a_n\rangle_G, \; \forall \, i = 1, \ldots, n \,. \tag{2.68}$$

Consequently, $|G\rangle = |0 \ldots 0\rangle_G$. Moreover, projectors on these vectors can be written as

$$_G|a_1 \ldots a_n\rangle\langle a_1 \ldots a_n|_G = \prod_{i=1}^{n} \frac{(-1)^{a_i} g_i + \mathbb{1}}{2} \,. \tag{2.69}$$

In the following, we will refer to states that are diagonal in a graph state basis as **graph-diagonal states**.

It is also useful to know that the application of Z_i to $|a_1 \ldots a_n\rangle_G$ flips the ith bit, i.e.

$$Z_i |a_1 \ldots a_i \ldots a_n\rangle_G = |a_1 \ldots a_i \oplus 1 \ldots a_n\rangle_G \,. \tag{2.70}$$

Moreover, any state can be transformed into a graph-diagonal state by local transformations [76]. This can be done for any given graph in the following way: Given a state q, we can write it in the graph basis associated to the given graph as

$$\varrho = \sum \lambda_{i_1 i_2 \ldots i_n, j_1 j_2 \ldots j_n} \,_G|i_1 i_2 \ldots i_n\rangle\langle j_1 j_2 \ldots j_n|_G \,. \tag{2.71}$$

Here, the sum runs over all binary strings $i_1 i_2 \ldots i_n \in \{0,1\}^n$ and the same for $j_1 j_2 \ldots j_n$. Then, one can throw a coin and with probability $1/2$ leave the state untouched and in the other case apply g_1 to

2.5 Graph states

the state. Note that the latter is a local operation as it is a tensor product of Pauli operators. Through this stochastic operation, the state is changed into

$$\varrho \mapsto \tilde{\varrho} = \frac{1}{2}(\varrho + g_1 \varrho g_1^\dagger) \tag{2.72}$$

Note that $\tilde{\varrho}$ does not have any off-diagonal elements of the form ${}_G\langle i_1 i_2 \ldots i_n \rangle \langle j_1 j_2 \ldots j_n |_G$ with $i_1 \neq j_1$ anymore, as

$$g_1 \, {}_G|i_1 i_2 \ldots i_n\rangle\langle j_1 j_2 \ldots j_n|_G \, g_1^\dagger = - {}_G|i_1 i_2 \ldots i_n\rangle\langle j_1 j_2 \ldots j_n|_G \,, \tag{2.73}$$

if $i_1 \neq j_1$. If one adds the untouched part to such terms, these off-diagonal terms cancel.

It is now straightforward to complete the transformation by adding $n-1$ more rounds in which, first, we apply g_2 to $\tilde{\varrho}$ with probability $1/2$, then g_3 to the resulting state with probability $1/2$ etc. This finally results in a graph-diagonal state which can only be entangled if the original state was already entangled, as the described local operations cannot create entanglement.

2.5.3 Local unitary operations on graph states

An important class of operations are the **local unitary operations**, as these operations do not change the entanglement properties of a state and leave its physical properties unchanged. Two states $|\psi\rangle$ and $|\phi\rangle$ that are related via local unitaries $U_1, \ldots, U_n$,

$$|\psi\rangle = U_1 \otimes \cdots \otimes U_n |\phi\rangle \,, \tag{2.74}$$

are called **LU-equivalent**.

In Refs [77,78], it was shown that two graph states belonging to two different mathematical graphs can be LU-equivalent (when also allowing for permutations of qubits). In particular, two graphs that can be transformed into each other by a **local complementation** are LU-equivalent. A local complementation on a vertex i is an inversion of its neighborhood graph, i.e. one considers all qubits that are neighbors of i (but not i itself) and deletes all edges that connect these qubits with each other. Also, an edge is added between each two qubits in this neighborhood that are not connected.

For example, a local complementation on qubit 1 in star graph No. 9 of Fig. 2.2 results in the fully connected graph, in which each of the six vertices is connected with every other vertex. Therefore, the states associated to these graphs are LU-equivalent. Note that both graphs are LU-equivalent to the GHZ state as in Eq. (2.43) of six qubits.

It has been shown that, when taking into account states of up to six qubits, there are 19 LU-equivalence classes of connected graph states [77]. Note that the equivalence classes of up to eight qubits have been characterized in Ref. [79]. Figure 2.2 shows one representative state of each LU-equivalence class. Any graph state of six or less qubits can therefore be mapped by local unitaries and permutations onto a state associated to some graph in Fig. 2.2. In terms of a graph transformation, these local unitaries all correspond to local complementations as described before.

The application of these local unitaries, whose exact form is given in Ref. [77], allows one to transform a witness for any graph state in a particular LU-equivalence class into a witness of any other graph state in the same class. However, in order to perform this transformation, it is important to know how the generators of the graph transform under local complementation. Consider a graph G and its generators g_i according to Eq. (2.57). Moreover, let $\widetilde{G}$ be the graph that is obtained from G by a local complementation on vertex i and $\widetilde{g}_i$ its generators according to Eq. (2.57). Then, this local complementation maps

$$g_j \mapsto \begin{cases} \widetilde{g}_j & \text{if } j \notin \mathcal{N}(i) \\ \widetilde{g}_j \widetilde{g}_i & \text{if } j \in \mathcal{N}(i) \end{cases}. \tag{2.75}$$

One can check that one can build up all possible products of generators $\widetilde{g}_i$ using the operators on the right-hand side of Eq. (2.75). Therefore, this mapping maps

$$|G\rangle\langle G| = \sum_{\vec{x}\in\{0,1\}^n} \prod_{i=1}^n g_i^{x_i} \mapsto \sum_{\vec{x}\in\{0,1\}^n} \prod_{i=1}^n \widetilde{g}_i^{x_i} = |\widetilde{G}\rangle\langle \widetilde{G}|. \tag{2.76}$$

as desired. An example of Eq. (2.75) will be given in Sec. 5.2.1, where the mapping will be applied to witness operators that we will construct there.

Finally, we add that for any graph state $|G\rangle$, the operator $W_{\text{proj}} = \frac{1}{2}\mathbb{1} - |G\rangle\langle G|$ is a witness [72]. We will refer to W_{proj} as the **projector witness** of $|G\rangle$ [80].

2.6 Semidefinite programming

To conclude the introduction, the reader will be introduced to a class of problems that can be solved numerically in an efficient way which will be presented in Sec. 2.6.1. Moreover, these problems allow for statements of existence and uniqueness which will be given in Sec. 2.6.2.

2.6.1 General form of a semidefinite program

Semidefinite programs have proven to be very useful in quantum information. The general form of a semidefinite program is given in the following definition [81].

Definition 17. *Given a vector $\vec{c} \in \mathbb{R}^m$ and a set of $m+1$ hermitian matrices $\{F_0, F_1, \ldots, F_m \mid F_i \in \mathbb{C}^{n\times n}\}$. Then, a minimization over $\vec{x} \in \mathbb{R}^m$ of the form*

2.6 Semidefinite programming

$$\min_{\vec{x}} \vec{c}\,\vec{x}$$

$$\text{subject to } F_0 + \sum_{i=1}^{m} x_i F_i \geq 0 \qquad (2.77)$$

$$(2.78)$$

is a **semidefinite program** (SDP). Note that the second line means that the matrix $F(\vec{x}) = F_0 + \sum_{i=1}^{m} x_i F_i$ should be positive semidefinite.

The set over which the optimization is performed is convex. This can be seen by noting that if $F(\vec{x}) \geq 0$ and $F(\vec{y}) \geq 0$ for some $x, y \in \mathbb{R}^m$, the linearity of F implies that also

$$F(t\vec{x} + (1-t)\vec{y}) = tF(\vec{x}) + (1-t)F(\vec{y}) \geq 0 \qquad (2.79)$$

for any t with $0 \leq t \leq 1$. Moreover, the function $g(\vec{x}) = \vec{c}\,\vec{x}$ which is optimized is convex as it is linear. An optimization of a convex function over a convex set is called a **convex optimization problem** and an SDP is a special case of such a problem, as just argued.

An example for an SDP would be to calculate the sum of all negative eigenvalues of a given Hermitian matrix A, which could be formulated as

$$\min_{P} \text{Tr}(AP)$$

$$\text{subject to } \mathbb{1} \geq P \geq 0\,. \qquad (2.80)$$

Here, $\mathbb{1} \geq P$ denotes the condition that $\mathbb{1} - P$ is positive semidefinite. This minimization is an SDP, where the vector $\vec{c}$ of Eq. (2.77) corresponds to the vector $\vec{p}$ in which all columns of P have been concatenated. Then, $\text{Tr}(AP) = \sum_i A_{ii} P_{ii}$ can be written as a scalar product of a vector $\vec{p}$ and a vector $\vec{c}$ that contains the diagonal elements A_{ii} and zeros for the non-diagonal elements of P. Moreover, the condition that P (and $\mathbb{1} - P$) must be positive semidefinite, can clearly be written as in Eq. (2.77) by choosing the matrices F_i such that $\sum_i p_i F_i = P$. This is not difficult, as p_i are the elements of $\vec{p}$ and therefore of the columns of P.

The minimum in Eq. (2.80) would be obtained for $P = \sum_i |e_i\rangle\langle e_i|$, where $|e_i\rangle$ are the eigenvectors of A with negative eigenvalues. In case A has no negative eigenvalues, the minimum is zero and obtained for $P = 0$. The SDP of Eq. (2.80) can also be used to calculate the negativity (cf. Definition 14) of a state ϱ. In order to do so, one has to replace A by ϱ^{T_A}.

Moreover, if the matrices F_i of Eq. (2.77) are diagonal, their sum is also diagonal. Then, the positivity constraint becomes much easier, as only the element-wise positivity of the diagonal elements has to be checked. These special instances of SDPs are called linear programs and will play a role in Sec. 5.1.

25

Definition 18. *Given a vector $\vec{c} \in \mathbb{R}^m$, $\vec{b} \in \mathbb{R}^n$ and a real matrix $A \in \mathbb{R}^{n \times m}$. Then, a minimization over $\vec{x} \in \mathbb{R}^m$ of the form*

$$\min_{\vec{x}} \vec{c}\,\vec{x}$$
$$\text{subject to } A\vec{x} + \vec{b} \geq 0 \tag{2.81}$$
$$\tag{2.82}$$

*is a **linear program** (LP). Note that the second line denotes element-wise positivity here.*

It becomes clear that any LP is an SDP, if one denotes the m column vectors of A by $\vec{a}_i$ and sets $F_0 = \text{diag}(\vec{b})$ and $F_i = \text{diag}(\vec{a}_i)$. Here, $\text{diag}(\vec{b})$ denotes the diagonal matrix whose diagonal elements are the entries of $\vec{b}$. With this choice, the LP of Eq. (2.81) has been brought into the form of Eq. (2.77) and is therefore an SDP.

2.6.2 Properties of a semidefinite program

Every SDP also has a dual problem which is defined as follows.

Definition 19. *Given a semidefinite program as in Def. 17. Then, its **dual problem** is*

$$\max_{Z} -\text{Tr}(F_0 Z) \tag{2.83}$$
$$\text{subject to } \text{Tr}(F_i Z) = c_i, \; i = 1, \ldots, m$$
$$Z \geq 0$$

where $Z \in \mathbb{C}^{n \times n}$ is a hermitian matrix and $Z \geq 0$ denotes positivity of the matrix Z.

The dual problem can be cast into the form of Eq. (2.77) and is therefore also an SDP. Since SDPs with $\vec{c} = 0$ test whether there exists a $\vec{x}$ that fulfills the constraints, they are also called **feasibility problems** and if any solution $\vec{x}$ exists, the SDP is **primal feasible**. Analogously, if there is a hermitian, positive semidefinite matrix Z with $\text{Tr}(F_i Z) = c_i$, then the SDP is **dual feasible**.

If a problem is primal and dual feasible, any $\vec{x}$ and Z that fulfill the constraints of the primal and dual problem, respectively, fulfill

$$\vec{c}\,\vec{x} + \text{Tr}(Z\,F_0) = \text{Tr}(ZF(\vec{x})) \geq 0, \tag{2.84}$$

where the positivity follows from $Z \geq 0$, $F(\vec{x}) \geq 0$. The fact that

$$\vec{c}\,\vec{x} \geq -\text{Tr}(Z\,F_0) \tag{2.85}$$

which follows from Eq. (2.84) is called **weak duality**. If the infimum of the left-hand side of Eq. (2.85) coincides with the supremum of its right-hand side, once speaks about **strong duality**. The conditions for this case are given by the following a theorem.

2.6 Semidefinite programming

Theorem 20. *Given an SDP. Let $p^* = \inf\{\vec{c}\vec{x} \,|\, F(\vec{x}) \geq 0\}$ be the optimal value of the primal problem and $d^* = \sup\{-\operatorname{Tr}(F_0\, Z) \,|\, Z = Z^\dagger \geq 0,\ \operatorname{Tr}(F_i\, Z) = c_i \,\forall\, i = 1, \ldots, m\}$ be the optimal value of the dual problem. If ...*

(i) ... there exists an $\vec{x} \in \mathbb{R}^m$ in the primal problem, such that $F(\vec{x}) > 0$ or ...

(ii) ... there exists a hermitian $Z \in \mathbb{C}^{n \times n}$ with $\operatorname{Tr}(F_i\, Z) = c_i$ for $i = 1, \ldots, m$ and $Z > 0$, ...

... then $p^* = d^*$. If (i) and (ii) hold, then $p^* \in \{\vec{c}\vec{x} \,|\, F(\vec{x}) \geq 0\}$ and $d^* \in \{-\operatorname{Tr}(F_0\, Z) \,|\, Z = Z^\dagger \geq 0,\ \operatorname{Tr}(F_i\, Z) = c_i \,\forall\, i = 1, \ldots, m\}$.

Duality is a strong tool that allows for testing the quality of a numerically performed minimization in an SDP. Assume that p_num is a minimum of the primal problem which has been determined numerically, which will therefore be slightly larger than the exact minimum, $p_\text{num} > p^*$. Also, let $d_\text{num} < d^*$ be a numerical estimate for the maximum of the dual problem. Then, due to weak duality,

$$d_\text{num} < d^* \leq p^* < p_\text{num}\,. \tag{2.86}$$

Therefore, p_num and d_num provide an interval in which the exact minimum of the primal problem must be. In the case of strong duality, the less-or-equal-sign in Eq. (2.86) is replaced by equality. Therefore, the difference between p_num and d_num provides a good measure of how well the numerical optimization worked.

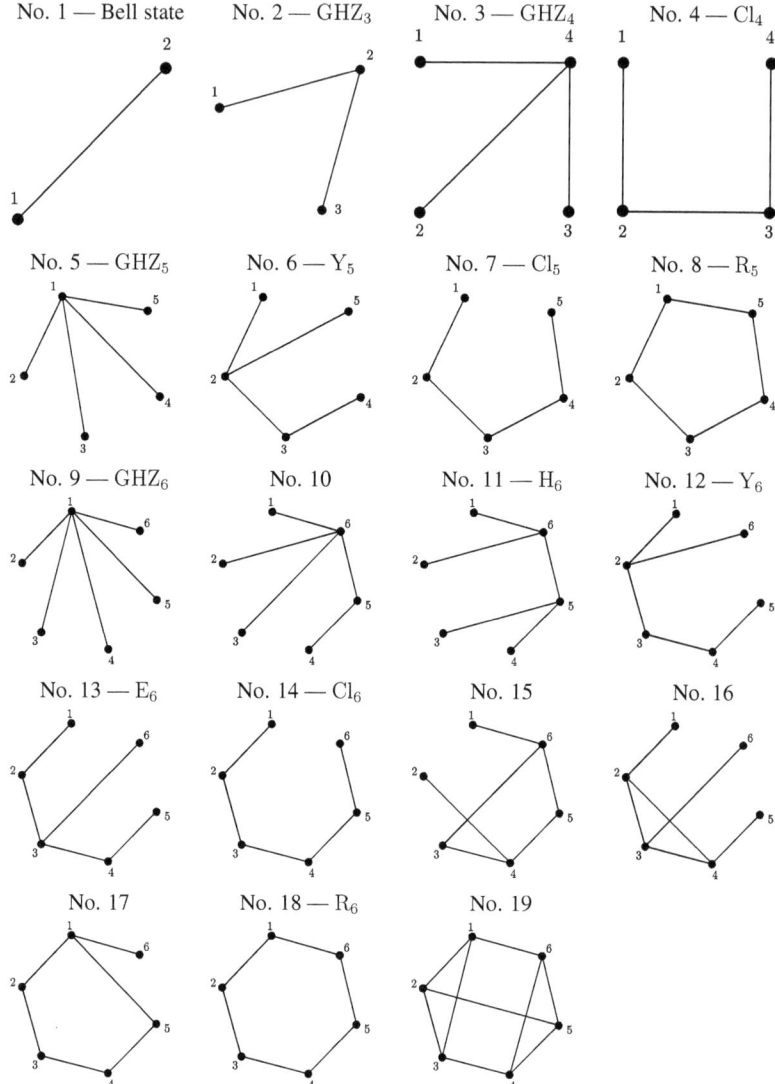

Figure 2.2: The graph states of up to six qubits can be grouped into 19 LU equivalence classes. For each class, we show the representative state here.

3 Bell inequalities: Statistical significances in experiments

In this chapter, we will consider an experimental test for entanglement and take its statistical error into account to ask how strong such a test really is. The main results of this section have already been published elsewhere.[1]

As quantum theory is a statistical theory, predicting in general only probabilities for experimental results, in most experiments observing quantum effects, several copies of a quantum state are generated and individually measured to determine the desired probabilities. As only a finite number of states can be generated, this leads to an unavoidable statistical error. The particularly low generation rate in certain experiments demands a careful statistical treatment.

For the experimental verification of entanglement, often inequalities for the correlations — such as Bell inequalities or entanglement witnesses (cf. Secs 2.2.2, 2.4) — are used, in which a violation indicates entanglement [31, 32]. The maximization of this violation has been investigated in detail, cf. Refs [19, 33]. In fact, making such inequalities more sensitive is a crucial step in order to allow advanced experiments with more particles.

In this chapter, it is demonstrated theoretically and experimentally that such an optimization does not necessarily lead to a better entanglement test, if the statistical nature of quantum theory is taken into account. It was already noted [83, 84] that, when aiming at ruling out local realism, highly entangled states do not necessarily deliver a stronger test than weakly entangled states, but this does not answer the question which inequality to use for a given state and it remains unclear how to apply it to actual error models used in experiments. Also, most of the different entanglement detection methods compared in Ref. [85] cannot be applied to multiparticle systems.

Theoretically, we show for different error models that decreasing the violation of an inequality can improve the significance. Also, we demonstrate this phenomenon in a four-photon experiment, measuring the Mermin and the Ardehali inequality. We find that the former inequality leads to a higher significance than the latter, despite a lower violation. Finally, we discuss the physical origin of this phenomenon and provide methods to construct entanglement tests with a high statistical significance.

---

[1] Reprinted excerpts with permission from Ref. [82], http://prl.aps.org/abstract/PRL/v104/i21/e210401. Copyright (2010) by the American Physical Society.

3 Bell inequalities: Statistical significances in experiments

3.1 Statement of the problem

Both Bell inequalities and entanglement witnesses provide a necessary condition for a state to be separable which can be formulated in the form of an inequality. An entanglement witness is positive on all separable states, while a Bell inequality

$$\langle \mathcal{B} \rangle \leq C_{\text{lhv}}, \qquad (3.1)$$

with some real constant C_{lhv} holds for all local hidden variable models (LHV models) and therefore for all separable states. In both cases, a violation implies entanglement and we define $\mathcal{V}$ as the violation of the corresponding inequality. That is, for a witness we have $\mathcal{V}(W) = -\langle W \rangle$ while for a Bell inequality $\mathcal{V}(\mathcal{B}) = \langle \mathcal{B} \rangle - C_{\text{lhv}}$. Then, the significance of an entanglement test can be defined as

$$\mathcal{S} = \frac{\mathcal{V}}{\mathcal{E}} \qquad (3.2)$$

where $\mathcal{E}$ is the statistical error for the experiment. Clearly, $\mathcal{E}$ depends on the particular experimental implementation and on the error model used. Nevertheless, in any experiment $\mathcal{S}$ is a well characterized quantity; its notion is widely used in the literature, when the violation is expressed in terms of "standard deviations", e.g. [25, 80].

Previously, much effort has been devoted to improving entanglement tests in order to achieve a higher violation. For instance, for entanglement witnesses a mature theory how to optimize witnesses has been developed [33]. Here, for a given witness W one tries to find a positive operator P, such that $W' = W - P$ is still a witness. In order to have a more significant result, however, one can either increase $\mathcal{V}$ in Eq. (3.2) or decrease $\mathcal{E}$. It is a central result of this chapter that decreasing $\mathcal{E}$ is often superior.

3.2 Variance as error

Let us first consider a simple model, in which we take the square root of the variance as the error of a witness,

$$\mathcal{E}(W) = \Delta(W) = \sqrt{\langle W^2 \rangle - \langle W \rangle^2}. \qquad (3.3)$$

An experimentally relevant model will be discussed below. This simple model already demonstrates that the standard optimization of witnesses is often not the appropriate approach to increase the significance:

Lemma 21. *Let $\varrho = |\psi\rangle\langle\psi|$ be a pure state detected by the witness W. Then, one can always increase the significance of W at the expense of optimality (i.e., by adding a positive operator). With this method one can make the significance arbitrarily large.*

Proof. We use as an ansatz for the improved witness $W' = W + \gamma P$, where $\gamma > 0$ and P is a positive observable with unit trace. For small γ, we expand

3.3 Error model for multi-photon experiments

$$-\frac{\langle W'\rangle}{\Delta(W')} = -\frac{\langle W\rangle}{\Delta(W)} + \gamma\frac{\langle W\rangle}{2\Delta^3(W)}\left(\langle WP+PW\rangle - 2\frac{\langle W^2\rangle}{\langle W\rangle}\langle P\rangle\right) + O(\gamma^2). \quad (3.4)$$

Maximizing this expression over all positive P with $\text{Tr}(P) = 1$ is equivalent to minimizing $\text{Tr}(QP)$, where $Q = \varrho W + W\varrho - 2\langle W^2\rangle/\langle W\rangle\varrho$. Hence the optimal P is a one-dimensional projector $P = |\varphi\rangle\langle\varphi|$, where $|\varphi\rangle$ is an eigenvector corresponding to the minimal eigenvalue of Q.

We still have to show that this minimal eigenvalue is negative. To this end, we make the ansatz $|\varphi\rangle = \alpha|\psi\rangle + \beta|\psi^\perp\rangle$, where $\langle\psi|\psi^\perp\rangle = 0$. We then have to minimize

$$\text{Tr}(QP) = 2\text{Re}(\alpha^\star\beta\langle\psi|W|\psi^\perp\rangle) - 2|\alpha|^2\frac{\Delta_\psi^2(W)}{\langle\psi|W|\psi\rangle}. \quad (3.5)$$

We can always choose the phases of α and β such that $\text{Re}(\alpha^\star\beta\langle\psi|W|\psi^\perp\rangle)$ is negative. Therefore the optimal $|\psi^\perp\rangle$ is the vector orthogonal to $|\psi\rangle$ which maximizes $|\langle\psi|W|\psi^\perp\rangle|$, i.e.,

$$|\psi_{\text{opt}}^\perp\rangle = [\mathbb{1} - |\psi\rangle\langle\psi|]W|\psi\rangle/\Delta_\psi(W). \quad (3.6)$$

Furthermore, we can always choose the moduli of α and β such that the negative term $2\text{Re}(\alpha^\star\beta\langle\psi|W|\psi^\perp\rangle)$ dominates the positive second term. This shows that the minimal eigenvalue of Q is negative.

For finite γ we can iterate this procedure. We always find the same $|\psi_{\text{opt}}^\perp\rangle$ (though α and β will be different in each iteration step). Thus, we make the ansatz

$$\gamma P = a|\psi\rangle\langle\psi| + b|\psi_{\text{opt}}^\perp\rangle\langle\psi_{\text{opt}}^\perp| + c|\psi\rangle\langle\psi_{\text{opt}}^\perp| + \text{h. c.} \quad (3.7)$$

for the final result of the iteration. If we choose $c = -\Delta_\psi(W)$, $ab \geq |c|^2$, and $a, b > 0$, then γP is positive, $|\psi\rangle$ is an eigenstate of W', and $\Delta_\psi(W')$ is zero, so $\mathcal{S}$ diverges. □

3.3 Error model for multi-photon experiments

Let us now consider a realistic situation, in which other and more specific error models are used. As our later implementation uses multi-photon entanglement, we concentrate on this type of experiments but our ideas can also be applied to other implementations, such as trapped ions.

The basic experimental quantities are the numbers of detection events n_i of the different detectors i. From these data, all other quantities such as correlations or mean values of observables are derived.

In the standard error model for photonic experiments [66, 86], the counts are assumed to be distributed according to a Poissonian distribution, whose mean value is given by the observed value. That is, for a certain measurement outcome i one sets the mean value as $\langle n_i\rangle = n_i$ and the error as $\mathcal{E}(n_i) = \sqrt{n_i}$ (being the standard deviation of a Poissonian distribution). In general, for a function $f = f(n_i)$ of several counts, Gaussian error propagation is applied to obtain the error (see below).

3 Bell inequalities: Statistical significances in experiments

To give an example, consider a two-qubit correlation

$$\mathcal{M} = \alpha Z_1 Z_2 + \beta Z_1 \mathbb{1}_2 + \gamma \mathbb{1}_1 Z_2 \,. \tag{3.8}$$

$\langle \mathcal{M} \rangle$ can be determined by measuring in the common eigenbasis of all three terms in $\mathcal{M}$, i.e., by projecting onto $|00\rangle, |01\rangle, |10\rangle$ and $|11\rangle$. Repeating this with many copies of the state will lead to count numbers n_{kl} with $k, l = 0$ or 1 and to count rates

$$p_{kl} = n_{kl}/n_{\text{tot}}, \text{ where } n_{\text{tot}} = n_{00} + n_{01} + n_{10} + n_{11} \,. \tag{3.9}$$

n_{tot} is the total number of events. The mean value $\langle \mathcal{M} \rangle$ can be written as a linear combination of p_{kl}, namely

$$\begin{aligned}\langle \mathcal{M} \rangle &= \lambda_{00} p_{00} + \lambda_{01} p_{01} + \lambda_{10} p_{10} + \lambda_{11} p_{11} \\ &\text{with } \lambda_{00} = \alpha + \beta + \gamma, \ \lambda_{01} = -\alpha + \beta - \gamma, \\ &\lambda_{10} = -\alpha - \beta + \gamma, \ \lambda_{11} = \alpha - \beta - \gamma \,.\end{aligned} \tag{3.10}$$

Then, according to Gaussian error propagation, the squared error is given by

$$\mathcal{E}(\mathcal{M})^2 = \sum_{k,l} \left(\frac{\partial \langle \mathcal{M} \rangle}{\partial n_{kl}} \right)^2 \mathcal{E}(n_{kl})^2 = \sum_{k,l} \left(\frac{\lambda_{kl}}{n_{\text{tot}}} - \frac{\langle \mathcal{M} \rangle}{n_{\text{tot}}} \right)^2 n_{kl} \,. \tag{3.11}$$

Using Eq. (3.9) and the definition of $\langle \mathcal{M} \rangle$ in Eq. (3.10), a simple calculation yields

$$\left(\frac{\lambda_{kl}}{n_{\text{tot}}} - \frac{\langle \mathcal{M} \rangle}{n_{\text{tot}}} \right)^2 n_{kl} = 4 \frac{n_{\pm}^2}{n_{\text{tot}}^4} n_{kl} \,, \tag{3.12}$$

where $n_{\pm}$ stands for n_-, the total number of counts associated to a negative eigenvalue, if $\lambda_{kl} = +1$, and for n_+, the total number of counts associated to a negative eigenvalue, if $\lambda_{kl} = -1$. Note that, in this example with two qubits, n_{00} and n_{11} are associated to a positive eigenvalue and therefore $n_+ = n_{00} + n_{11}$. Thus, $n_- = n_{01} + n_{10}$.

Now, when executing the sum in Eq. (3.11), one needs to distinguish between count numbers n_{kl} associated to negative eigenvalues and those associated to positive eigenvalues. Using that $n_+ + n_- = n_{\text{tot}}$, we arrive at

$$\mathcal{E}(\mathcal{M})^2 = \sum_{k,l} \left(\frac{\lambda_{kl}}{n_{\text{tot}}} - \frac{\langle \mathcal{M} \rangle}{n_{\text{tot}}} \right)^2 n_{kl} = 4 \frac{n_+ n_-}{n_{\text{tot}}^3} = 4 \frac{n_+ (n_{\text{tot}} - n_+)}{n_{\text{tot}}^3} \,. \tag{3.13}$$

This easy formula also shows that the error of a single setting is maximal if one measures an equal number of outcomes for $+1$ and for -1.

3.3 Error model for multi-photon experiments

In an experiment, one usually continues measuring until one has collected a certain amount of experimental data, i.e. a total number of N counts. In photonic experiments, however, measurements can usually only be performed locally on single qubits. In general, the observable to be measured contains, say, m terms that need to be measured in different local bases (one of which might be $\mathcal{M}$ of Eq. (3.8)). Now, one can ask how much time should be spent on measuring each setting in order to minimize the error.

Lemma 22. *Consider a photonic experiment in which the expectation value of an operator*

$$\mathcal{B} = \sum_{i=1}^{m} \alpha_i \mathcal{M}_i \qquad (3.14)$$

is to be determined. Here, the measurement of each term $\mathcal{M}_i$ has to be performed in a different basis. Assume that basis i, $1 \leq i \leq m$, is measured until one has detected $n_{\text{tot}}^{(i)}$ counts in this basis. Let $N = \sum_{i=1}^{m} n_{\text{tot}}^{(i)}$. Then, the total error is minimal if

$$n_{\text{tot}}^{(i)} = \frac{N |\alpha_i| |\mathcal{E}_i \left(\frac{N}{m}\right)|}{\sum\limits_{j=1}^{m} |\alpha_j| |\mathcal{E}_j \left(\frac{N}{m}\right)|}, \qquad (3.15)$$

where $\mathcal{E}_i \left(\frac{N}{m}\right)$ is the error of setting i in the case where each setting is measured equally often.

Corollary 23. *We denote the experimental state by $\varrho_{\exp}$. If both $|\alpha_i|$ and $|\text{Tr}(\mathcal{M}_i \varrho_{\exp})|$ are the same for each setting i, then the uniform distribution $n_{\text{tot}}^{(i)} = N/m$ achieves the minimal error.*

Proof. Lemma 22 — We need to minimize the total error, whose square is, according to Gaussian error propagation, given by

$$\mathcal{E}^2(n_{\text{tot}}^{(1)}, \ldots, n_{\text{tot}}^{(m)}) = \sum_{i=1}^{m} \alpha_i^2 \mathcal{E}_i^2(n_{\text{tot}}^{(i)}). \qquad (3.16)$$

Note that minimizing the squared error is equivalent with minimizing the error itself. We define x_i by $n_{\text{tot}}^{(i)} = x_i \frac{N}{m}$, so that it gives the factor by which $n_{\text{tot}}^{(i)}$ deviates from an equal distribution of the total number N of counts on the m different settings.

Now, $\mathcal{E}^2$ must be minimized under the condition that $N = \sum_{i=1}^{m}$, i.e. that $\sum_{i=1}^{m} x_i = m$. This can be done using a Lagrange multiplier λ by minimizing

$$f(x_1, \ldots, x_n, \lambda) = \mathcal{E}^2(n_{\text{tot}}^{(1)}, \ldots, n_{\text{tot}}^{(m)}) + \lambda(\sum_{i=1}^{m} x_i - m). \qquad (3.17)$$

Equation (3.13) shows that $\mathcal{E}_i^2\left(x_i \frac{N}{m}\right) = \frac{1}{x_i}\mathcal{E}_i^2\left(\frac{N}{m}\right)$ and therefore

$$f(x_1, ..., x_n, \lambda) = \sum_{i=1}^{m} \frac{\alpha_i^2}{x_i} \mathcal{E}_i^2\left(\frac{N}{m}\right) + \lambda(\sum_{i=1}^{m} x_i - m) \,. \tag{3.18}$$

This expression is minimal for the count numbers given in Eq. (3.15).

Corollary 23 — The corollary can be seen as follows: For any setting i, we have

$$n_+ n_- = \left(n_{\text{tot}}^{(i)}\right)^2 \frac{1}{4} [1 - \text{Tr}(\mathcal{M}_i \varrho_{\text{exp}})][1 + \text{Tr}(\mathcal{M}_i \varrho_{\text{exp}})] \tag{3.19}$$

$$= \left(n_{\text{tot}}^{(i)}\right)^2 \frac{1}{4} \{1 - [\text{Tr}(\mathcal{M}_i \varrho_{\text{exp}})]^2\} \,. \tag{3.20}$$

Therefore, if $|\text{Tr}(\mathcal{M}_i \varrho_{\text{exp}})|$ and $|\alpha_i|$ are setting-independent, the product $n_+ n_-$ is the same for each setting. According to Eq. (3.13) and Eq. (3.15), the optimal particle numbers are then the same for each setting and therefore equal the uniformly distributed ones. □

From now on, we will denote the count numbers n_{tot} in Eq. (3.13) by $n_{\text{tot}}^{(i)}$ as they only refer to the single setting i.

Let us finally discuss the underlying assumptions of this error model. The first main assumption is that the n_{kl} are Poisson distributed and their errors are uncorrelated.

This is well motivated by the experimental observations and assumes that the detector efficiency is low. Moreover, Gaussian error propagation stems from a Taylor expansion of the function f and therefore assumes that the non-linear terms of the expansion are small. Finally, one can use the standard deviation to define a confidence interval. Under the assumption that the distribution is Gaussian, one can state that, when repeating the measurement, with a probability of 68.27% the distance of the measurement outcome to the mean value will be smaller than one standard deviation. For other distributions the connection between confidence interval and standard deviation is not so direct. If the number of events for all detectors is sufficiently large (e.g. $n_{kl} \gtrsim 10$), however, the Poissonian distribution is approximated well by a Gaussian distribution.

3.4 Bell inequalities for four particles

We now consider the Mermin and Ardehali inequality for four qubits (cf. Sec. 2.4.2, [54, 56]). In order to be consistent with the notation of Eq. (3.1), we define [cf. Eqs (2.42) and Eqs (2.44)]

$$\begin{aligned}\mathcal{B}_M = {}& X_1 X_2 X_3 X_4 - X_1 X_2 Y_3 Y_4 - X_1 Y_2 Y_3 X_4 - Y_1 Y_2 X_3 X_4 \\ & - Y_1 X_2 Y_3 X_4 - X_1 Y_2 X_3 Y_4 - Y_1 X_2 X_3 Y_4 + Y_1 Y_2 Y_3 Y_4 \,,\end{aligned} \tag{3.21}$$

3.4 Bell inequalities for four particles

for the Mermin inequality [54] and

$$\begin{aligned}\mathcal{B}_A = \Big(&A_1X_2X_3X_4 + B_1X_2X_3X_4 - A_1X_2Y_3Y_4 - B_1X_2Y_3Y_4 \\&- A_1Y_2Y_3X_4 - B_1Y_2Y_3X_4 - A_1Y_2X_3Y_4 - B_1Y_2X_3Y_4 \\&- A_1Y_2X_3X_4 + B_1Y_2X_3X_4 - A_1X_2X_3Y_4 + B_1X_2X_3Y_4 \\&- A_1X_2Y_3X_4 + B_1X_2Y_3X_4 + A_1Y_2Y_3Y_4 - B_1Y_2Y_3Y_4\Big)/\sqrt{2}\,,\end{aligned} \quad (3.22)$$

for the Ardehali inequality [56]. Here, $A_1 = (X_1 + Y_1)/\sqrt{2}$ and $B_1 = (X_1 - Y_1)/\sqrt{2}$. We wrote $\mathcal{B}_M$ and $\mathcal{B}_A$ with the Pauli matrices as observables, since they are used later, however, one might replace them by arbitrary dichotomic measurements.

The Mermin and Ardehali inequality reveal the non-local correlations of the four-qubit GHZ state $|GHZ_4\rangle$ [cf. Eq. (2.43)]. For this state we have $\langle\mathcal{B}_M\rangle = \langle\mathcal{B}_A\rangle = 8$. As the bound for LHV models for the Ardehali inequality is smaller, the violation $\mathcal{V}$ is larger. This may lead to the opinion that the Ardehali inequality is "better" than the Mermin inequality for the state $|GHZ_4\rangle$.

However, this belief is easily shattered, if the significance $\mathcal{S}$ is considered as the relevant figure of merit. This can be seen directly from Eq. (3.11). The GHZ state is an eigenstate for each of the correlation measurements in the Mermin inequality (they are stabilizing operators of the GHZ state). Hence, if the Mermin inequality for a perfect GHZ state is measured, we have in the last term of Eq. (3.11) for each case k, l either $\lambda_{kl} = \langle\mathcal{M}\rangle$ (since the mean value is an eigenvalue) or $n_{kl} = 0$, hence $\mathcal{E}(\mathcal{M})$ vanishes. The Ardehali inequality, however, does not contain stabilizer terms and the error remains finite.

For an experimental application it is important that the Mermin inequality leads to a higher significance than the Ardehali inequality, even if noise is introduced. [2] To see this, we considered bit-flip noise, which can easily be simulated in experiment. Therefore, we used a perfect GHZ state whose qubits are locally affected by the bit-flip operation f with probability p, i.e.

$$f(\varrho_i) = (1-p)\varrho_i + pX_i\varrho_iX_i\,, \text{ for each qubit } i\,. \quad (3.23)$$

In Fig. 3.1 a), we plotted the significance $\mathcal{S}$ versus the fidelity F of the noisy state w.r.t. a perfect GHZ state, i.e. $F = \langle GHZ_4|\varrho_{exp}|GHZ_4\rangle$, and versus the bit-flip probability p. We considered the case in which each setting is measured the same number of times and therefore $n_{\text{tot}}^{(1)} = n_{\text{tot}}^{(2)} = \ldots$. For $F \geq 0.70$ the Mermin inequality is more significant (for the 6-qubit versions of these inequalities [54, 56], this changes to $F \geq 0.40$).

Moreover, it is possible to optimize the count numbers $n_{\text{tot}}^{(i)}$ according to Lemma 22 to minimize the error and therefore increase the statistical significance. Doing so results in the dotted curves in Fig. 3.1

---

[2] This is also important, as for nearly perfect GHZ states, some count numbers n_{kl} will be close to zero. Then, the interpretation of the statistical error as a confidence interval may be questioned. Note that a small count number n_- or a small n_+ also implies that the error is small according to Eq. (3.13) and hence the statistical significance high.

3 Bell inequalities: Statistical significances in experiments

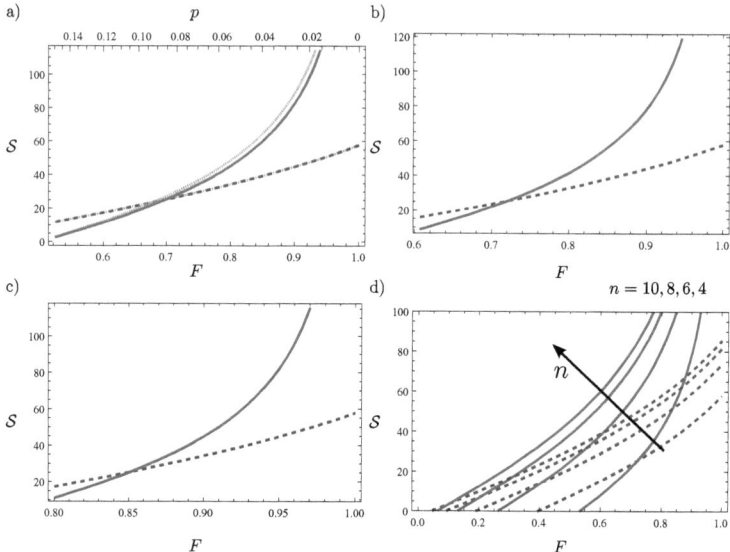

Figure 3.1: a) Significance $\mathcal{S}$ for the Mermin (solid curve) and the Ardehali inequality (dashed curve) for bit-flip noise when measuring each setting equally often. Using optimized count numbers (cf. Lemma 22) results in the dotted curves. Note that for the Ardehali inequality, the dotted and dashed curves coincide. On the horizontal axes, we show the bit-flip probability and the corresponding fidelity with respect to a perfect GHZ state. We assumed that the experimenter prepares 8000 instances of a GHZ state and chooses either to measure the eight terms of the Mermin inequality (each term with 1000 realizations of the state) or the 16 terms of the Ardehali inequality with 500 states per correlation term. See text for further details. In b), the same situation but with white noise added to the GHZ state was considered. c) shows dephasing noise. d) shows the Mermin (solid) and the Ardehali inequality (dashed) for white noise for an increasing number of qubits n. The values plotted here are $n = 4, 6, 8, 10$ and the arrow points in the direction of increasing n.

a). One can see that, while the significance of the Mermin inequality can be enhanced considerable, the significance of the Ardehali inequality does not change considerably. In fact, it only increases by a factor of approximately 1.003 for $F = 0.6$ and even less at higher fidelities.

The sharp-eyed reader might have noticed that Eq. (3.15) results in vanishing count numbers for settings i for which the error $\mathcal{E}_i(\frac{N}{m})$ is zero. To produce the dotted curves in Fig. 3.1 a), we therefore assumed that in such cases, for setting i one count would be measured. Since for $F = 1$ the total error

3.4 Bell inequalities for four particles

already vanishes for a uniform distribution in the count numbers, an application of Eq. (3.15) cannot decrease the error anymore and is, indeed, impossible as its denominator vanishes.

As can be seen from Eq. (3.11), the fact that one witness is more significant than the other one is independent of the total count number $N = \sum_{i=1}^{m} n_{\text{tot}}^{(i)}$. This holds both for the case in which all $n_{\text{tot}}^{(i)}$ are the same and in the case where they obey Eq. (3.15). This behavior is caused by the fact that n_{kl} is proportional to $n_{\text{tot}}^{(i)}$, which is proportional to N and therefore the intersection point of both curves is independent from N.

Moreover, although not implemented in our experiment, let us examine what happens for white noise. In fact, a calculation yields very similar values ($F \geq 0.72$ for 4 qubits, $F \geq 0.41$ for 6 qubits). The situation of white noise and four qubits is shown in Fig. 3.1 b), while c) shows dephasing noise ($F \geq 0.85$ for 4 qubits), i.e. a decay of the off-diagonal matrix elements of the GHZ state.

Figures 3.1 a), b), c) suggest that the qualitative behavior of the significances does not depend on details of the noise; only the scaling of the fidelity axis changes. For white noise, the following statement can be made:

Lemma 24. *For a state*

$$\varrho(p) = (1-p)|GHZ_n\rangle\langle GHZ_n| + \frac{p}{2^n}\mathbb{1}, \qquad (3.24)$$

the statistical significance of the Mermin inequality for an even number of particles n is given by

$$S_M(p) = \sqrt{N}\frac{\frac{1}{2}(1-p) - 2^{-n/2}}{\sqrt{\frac{p}{2}(1-\frac{p}{2})}}, \qquad (3.25)$$

when one measures each term in the inequality the same number of time. For an odd number of qubits, $2^{-n/2}$ is replaced by $2^{-(n+1)/2}$. The statistical significance of the Ardehali inequality is, for even n, given by

$$S_A(p) = \sqrt{N}\frac{\frac{1}{2}(1-p) - 2^{-(n+1)/2}}{\sqrt{\frac{p}{2} - \frac{p^2}{4} + \frac{1}{4}}}. \qquad (3.26)$$

For odd n, $2^{-(n+1)/2}$ is replaced by $2^{-n/2}$. Here, N is the total number of photons detected (summed over all different settings).

It turns out that the range of values of p, in which the Mermin inequality is more significant, becomes even larger for an increasing particle numbers. In fact, the following statement describes its behavior.

Corollary 25. *For increasing particle number, the range of values of p and therefore also the range of values of $F = \langle GHZ_n|\varrho(p)|GHZ_n\rangle$, for which the Mermin inequality is better than the Ardehali inequality increases exponentially [cf. Fig. 3.1 d)]. In other words, the value of p, from which on the Mermin inequality is worse, converges to one exponentially fast.*

Proof. Mermin inequality —Let us prove Lemma 24 first for the Mermin inequality. Note that the

$\mathcal{B}_M$ can be written as

$$\mathcal{B}_M = \sum_{i=1}^{2^{n-1}} m_i M_i = \sum_{i=1}^{2^{n-1}} m_i(M_i^+ - M_i^-), \tag{3.27}$$

where,

$$\text{for any } i \in \{1, \ldots, 2^{n-1}\} : m_i = \pm 1 \text{ and } M_i|GHZ_n\rangle = \mp|GHZ_n\rangle. \tag{3.28}$$

M_i are tensor products of Pauli matrices and can be written as a difference of M_i^+, M^-, where M_i^+ is a projector onto the positive eigenspace of M_i and M_i^- a projector onto the negative eigenspace.

Then, the number of positive counts n_i^+ when measuring M_i for a state as in Eq. (3.24) is given by

$$n_i^+ = n_i^{\text{tot}} \left[\langle GHZ_3|M_i^+|GHZ_3\rangle (1-p) + \frac{p}{2} \right], \tag{3.29}$$

where n_i^{tot} is the total count number when measuring M_i. Note that $\text{Tr}(M_i^+) = 2^{n-1}$ was used here. Due to Eq. (3.28),

$$\delta_i = \langle GHZ_3|M_i^+|GHZ_3\rangle \in \{0,1\}. \tag{3.30}$$

When we use Eq. (3.13) for all settings i and plug in Eqs. (3.29) and (3.29), we then obtain

$$\mathcal{E}^2 = \sum_{i=1}^{2^{n-1}} \frac{4}{n_i^{\text{tot}}} \left[1 - \delta_i(1-p) - \frac{p}{2} \right] \left[\delta_i(1-p) + \frac{p}{2} \right] \tag{3.31}$$

$$= \sum_{i=1}^{2^{n-1}} \frac{4}{n_i^{\text{tot}}} \left[\delta_i(1-p) + \frac{p}{2} - \delta_i^2(1-p)^2 - \delta_i(1-p)p - \frac{p^2}{4} \right] \tag{3.32}$$

$$= \sum_{i=1}^{2^{n-1}} \frac{4}{n_i^{\text{tot}}} \left[(\delta_i - \delta_i^2)(1-p)^2 + \frac{p}{2} - \frac{p^2}{4} \right] \tag{3.33}$$

$$= \sum_{i=1}^{2^{n-1}} \frac{4}{n_i^{\text{tot}}} \left(\frac{p}{2} - \frac{p^2}{4} \right) \tag{3.34}$$

$$= 2^{2n} \frac{1}{N} \frac{p}{2} (1 - \frac{p}{2}). \tag{3.35}$$

Here, we have used that $\delta_i^2 = \delta_i$ in the fourth line and, in the last line, the assumption that each setting is measured equally often and therefore $n_i^{\text{tot}} = N/2^{n-1}$.

A straightforward calculation shows that the violation of the Mermin inequality for $\varrho(p)$ is given by

$$\mathcal{V}_M = \begin{cases} 2^{n-1}(1-p) - 2^{n/2}, & \text{if } n \text{ is even} \\ 2^{n-1}(1-p) - 2^{(n-1)/2}, & \text{if } n \text{ is odd} \end{cases}. \tag{3.36}$$

3.4 Bell inequalities for four particles

Thus, the statistical significance for the Mermin inequality for an even number of qubits is given by

$$\mathcal{S}_M = \mathcal{V}_M/\mathcal{E} = \frac{\sqrt{N}}{2^n} \frac{2^{n-1}(1-p) - 2^{n/2}}{\sqrt{\frac{p}{2}(1-\frac{p}{2})}}.\tag{3.37}$$

For an odd number of qubits, $2^{n/2}$ is replaced by $2^{(n-1)/2}$.

Ardehali inequality — Analogously to the last part of the proof, we write the Ardehali inequality as

$$\mathcal{B}_A = \frac{1}{\sqrt{2}} \sum_{i=1}^{2^{n-1}} \alpha_i A_i = \frac{1}{\sqrt{2}} \sum_{i=1}^{2^{n-1}} \alpha_i (A_i^+ - A_i^-),\tag{3.38}$$

where A_i are tensor products of local operators, A_i^+ is a projector onto the positive eigenspace of A_i and A_i^- a projector onto the negative eigenspace. The coefficients α_i take values ± 1.

Again, the number of positive counts n_i^+ when measuring A_i for a state as in Eq. (3.24) is given by

$$n_i^+ = n_i^{\text{tot}} \left[\langle GHZ_3|A_i^+|GHZ_3\rangle (1-p) + \frac{p}{2} \right].\tag{3.39}$$

We define $\alpha_i = \langle GHZ_3|A_i^+|GHZ_3\rangle$ and it remains to calculate these coefficients. We note that

$$\langle GHZ_n|Y_1 Y_2 \ldots Y_m X_{m+1} \ldots X_n|GHZ_n\rangle = \begin{cases} +1, & \text{if } m/4 \text{ is an integer} \\ -1, & \text{if } m \text{ is even, but } m/4 \text{ not an integer} \\ 0, & \text{if } m \text{ is odd} \end{cases}\tag{3.40}$$

holds and is also valid for any permutation of the operator $Y_1 Y_2 \ldots Y_m X_{m+1} \ldots X_n$. Note that every A_i is the sum of two terms of the form as in Eq. (3.40) (with an additional factor of $1/\sqrt{2}$. One of these terms includes X_1, while the other one includes one more Y-operator, namely Y_1. Therefore, we have

$$\alpha_i = \langle GHZ_3|A_i^+|GHZ_3\rangle = \frac{1}{2}\langle GHZ_3|\mathbb{1} + A_i|GHZ_3\rangle = \frac{1}{2}(1 \pm \frac{1}{\sqrt{2}}).\tag{3.41}$$

Therefore,

$$\alpha_i - \alpha_i^2 = \frac{1}{2}(1 \pm \frac{1}{\sqrt{2}}) - \frac{1}{4}(\frac{3}{2} \pm \frac{2}{\sqrt{2}}) = \frac{1}{8}.\tag{3.42}$$

3 Bell inequalities: Statistical significances in experiments

Then, we have

$$\mathcal{E}^2 = \sum_{i=1}^{2^{n-1}} \frac{4}{n_i^{\text{tot}}} \left[(\alpha_i - \alpha_i^2)(1-p)^2 + \frac{p}{2} - \frac{p^2}{4} \right] \quad (3.43)$$

$$= \sum_{i=1}^{2^{n-1}} \frac{4}{n_i^{\text{tot}}} \left[\frac{1}{8}(1-p)^2 + \frac{p}{2} - \frac{p^2}{4} \right] \quad (3.44)$$

$$= \frac{2^{n+1}}{N} \left[(1-p)^2 2^{n-3} + 2^{n-1}(p - \frac{p^2}{2}) \right] \quad (3.45)$$

$$= 2^{2n} \frac{1}{N} \left(\frac{p}{2} - \frac{p^2}{4} + \frac{1}{4} \right). \quad (3.46)$$

Here, the first line is the same as in Eq. (3.31), but for the second line, we used Eq. (3.42). In the third line, the assumption that each term is measured the same number of times was used. The violation is then given by

$$\mathcal{V}_A = \begin{cases} 2^{n-1}(1-p) - 2^{(n-1)/2}, & \text{if } n \text{ is even} \\ 2^{n-1}(1-p) - 2^{n/2}, & \text{if } n \text{ is odd} \end{cases} \quad (3.47)$$

and thus, for an even number of qubits,

$$\mathcal{S}_A = \mathcal{V}_A / \mathcal{E} = \frac{\sqrt{N}}{2^n} \frac{2^{n-1}(1-p) - 2^{(n-1)/2}}{\sqrt{\frac{p}{2} - \frac{p^2}{4} + \frac{1}{4}}}. \quad (3.48)$$

For an odd number of qubits, $2^{(n-1)/2}$ is replaced by $2^{n/2}$.

Corollary — Let us now show Corollary 25. In our proof, we are only going to use the quantities $\mathcal{S}_M(p)$ of Eq. (3.25) and $\mathcal{S}_A(p)$ Eq. (3.26) to prove that the range of p in which $\mathcal{S}_M(p) > \mathcal{S}_A(p)$ holds increases exponentially with n. Note that Eqs. (3.25) and (3.26) are the significances in the case of an even number of particles. However, for an odd number of particles, the significance of the Mermin inequality is even larger than the expression in Eq. (3.25), since it is obtained through replacing $2^{-n/2}$ by $2^{-(n+1)/2}$. On the contrary, the significance for the Ardehali inequality and an odd number of particles is smaller than the expression given in Eq. (3.26). Therefore, it is enough to show $\mathcal{S}_M(p) > \mathcal{S}_A(p)$ using the expressions in Eqs. (3.25) and (3.26) for an arbitrary number of particles.

Let us first show that the curves for the two significances can only intersect at most once in the range $0 \leq p \leq 1$. For $p = 0$, $\mathcal{S}_M(p)$ diverges, while $\mathcal{S}_A(p)$ does not. For $0 < p \leq 1$, both significances are differentiable with respect to p and their derivatives are continuous, so we can simply show that $\frac{\partial}{\partial p}[\mathcal{S}_A(p) - \mathcal{S}_M(p)] > 0$, which implies that there can only be at most one value for p at which this difference vanishes and the significances equal each other.

3.4 Bell inequalities for four particles

Due to Eqs (3.25) and (3.26), the derivatives are given by

$$\frac{\partial}{\partial p}\mathcal{S}_A(p) = -\sqrt{N}\frac{2+2^{(1-n)/2}(p-1)}{[1-(p-2)p]^{3/2}} \tag{3.49}$$

$$\frac{\partial}{\partial p}\mathcal{S}_M(p) = -\sqrt{N}\frac{1+2^{1-n/2}(p-1)}{[-(p-2)p]^{3/2}}. \tag{3.50}$$

Positivity of $\frac{\partial}{\partial p}[\mathcal{S}_A(p) - \mathcal{S}_M(p)]$ is equivalent with

$$-f(p,n) + g(p) > 0, \tag{3.51}$$

where we defined

$$f(p,n) = \frac{1}{\sqrt{2}}\frac{2+2^{(1-n)/2}(p-1)}{1/\sqrt{2}+2^{(1-n)/2}(p-1)}, \tag{3.52}$$

$$g(p) = \left[1 + \frac{1}{(2-p)p}\right]^{3/2}. \tag{3.53}$$

A straightforward calculation shows that f is monotonously decreasing with n for $0 \leq p \leq 1$. Therefore, if we can show that Eq. (3.51) holds for any $p \in [0,1]$ and, say, $n = 4$, the positivity must hold for all $n \geq 4$ for any fixed value of p.

Here, we chose $n = 4$, since it is particularly easy to verify: Let us divide the interval $[0,1]$ of the values that p can take into the two intervals $[0, 1/2]$ and $[1/2, 1]$. Then, we have $f(0, n = 4) = 4 - \sqrt{2}/2 < (7/3)^{3/2} = g(1/2)$. Together with the fact that both g and f are monotonously decreasing with p, i.e. the fact that f is maximal at $p = 0$ and g is minimal at $p = 1/2$ in the first interval, this implies that that $g(p) > f(p, n = 4)$ in the first interval. The same reasoning works for the second interval, as $f(1/2, n = 4) = (16 - \sqrt{2})/6 < 2\sqrt{2} = g(1)$. Thus, we know that, indeed, Eq. (3.51) holds for $0 \leq p \leq 1$ and $n = 4$ and, since f is monotonously decreasing with n, for all $n \geq 4$.

Thus, we have shown that there is at most one intersection point of $\mathcal{S}_A(p)$ and $\mathcal{S}_M(p)$ between 0 and 1 (for $n \geq 4$). Note that one can also show this for $n \geq 2$, but the calculation for $n = 2$ is much more complex and we are only interested in the behavior of the significances for large n.

Finally, we will show that, at the point $p_0 = 1 - (2 + \sqrt{2})2^{-n/2}$, we have $\mathcal{S}_M(p_0) > \mathcal{S}_A(p_0)$ for all values of n. This is equivalent with

$$\sqrt{\frac{\frac{p_0}{2}(1-\frac{p_0}{2})+\frac{1}{4}}{\frac{p_0}{2}(1-\frac{p_0}{2})}} - \frac{\frac{1}{2}(1-p_0) - 2^{-(n+1)/2}}{\frac{1}{2}(1-p_0) - 2^{-n/2}} > 0. \tag{3.54}$$

If we plug the definition of p_0 in the second term, this simplifies to

$$h(p_0) - \sqrt{2} > 0, \tag{3.55}$$

where we have also introduced the definition

$$h(p_0) = \sqrt{\frac{\frac{p_0}{2}(1 - \frac{p_0}{2}) + \frac{1}{4}}{\frac{p_0}{2}(1 - \frac{p_0}{2})}} \qquad (3.56)$$

for the first term in Eq. (3.54). As p_0 depends on n, so does $h(p_0)$. Note that, for $n < 4$, p_0 is negative and therefore not a probability. However, for any larger n, the derivative $\frac{dh}{dn} = \frac{\partial h}{\partial p_0} \frac{\partial p_0}{\partial n}$ is strictly negative, since $\frac{\partial h}{\partial p_0} < 0$ and $\frac{\partial p_0}{\partial n} > 0$. As $h(p_0)$ does not have a pole in the region $0 < p_0 \leq 1$, its negative derivative implies that it is always larger than $\lim_{n \to \infty} h(p_0) = \sqrt{2}$, which shows that $\mathcal{S}_M(p_0) - \mathcal{S}_A(p_0) > 0$.

Since p_0 approaches 1 exponentially fast with increasing n, the range in which the Mermin inequality is better than the Ardehali inequality also increases exponentially.

In terms of the fidelity, one also has an exponential increase due to

$$F_0 = \langle GHZ_n | \varrho(p_0) | GHZ_n \rangle = \left(2 + \sqrt{2}\right)\left(2^{-n/2} - 2^{-3n/2}\right) + 2^{-n}. \qquad (3.57)$$

□

Finally, we note that for white noise and dephasing noise, applying Lemma 22 does not improve the significances. In order to see that this is true for white noise, i.e. for the state $\varrho(p)$ of Eq. (3.24), we note that

$$\mathrm{Tr}[M_i \varrho(p)] = \mathrm{Tr}(M_i | GHZ_4 \rangle \langle GHZ_4 |) = \pm 1 \qquad (3.58)$$

and

$$\mathrm{Tr}[A_i \varrho(p)] = \mathrm{Tr}(A_i | GHZ_4 \rangle \langle GHZ_4 |) = \pm \frac{1}{\sqrt{2}} \qquad (3.59)$$

hold for all settings i and therefore Corollary 23 applies.

For dephasing noise, the noisy state can be written as

$$\tilde{\varrho}(p) = \frac{1}{2}(|0000\rangle\langle 0000| + |1111\rangle\langle 1111|) + \frac{p}{2}(|0000\rangle\langle 1111| + |1111\rangle\langle 0000|) \qquad (3.60)$$

$$= |GHZ_4\rangle\langle GHZ_4| - \frac{1-p}{16}\mathcal{B}_M \qquad (3.61)$$

Therefore, due to the form of $\mathcal{B}_M$, we have

$$\mathrm{Tr}(M_i \tilde{\varrho}(p)) = \pm 1 - (\pm 1)(1 - p) \qquad (3.62)$$

and Corollary 23 can be used. As $\mathcal{B}_A = \mathcal{B}_M$ and $A_i | GHZ_4 \rangle = \pm \frac{1}{\sqrt{2}} | GHZ_4 \rangle$, the same holds for the Ardehali inequality.

Let us now come to the an experimental implementation to see whether the effect of a large statistical significance for the Mermin inequality at high fidelities also occurs in the experiment.

3.5 Experimental setup

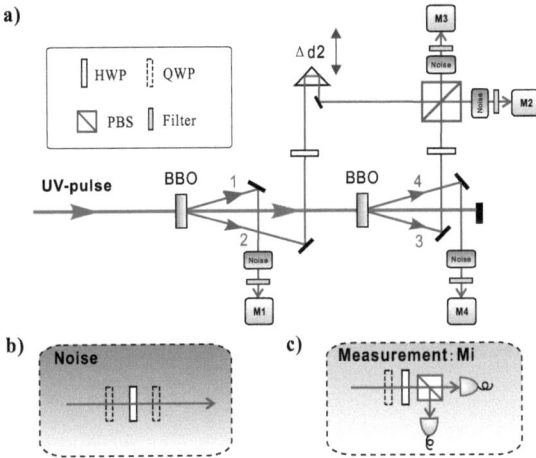

Figure 3.2: Scheme of the experimental setup. **a.** The setup to generate the required four-photon GHZ state. Femtosecond laser pulses ($\approx$ 200 fs, 76 MHz, 788 nm) are converted to ultraviolet pulses through a frequency doubler LiB_3O_5 (LBO) crystal (not shown). The pulses go through two main β-barium borate (BBO) crystals (2 mm), generating two pairs of photons. The observed two-fold coincidence count rates are about 1.6×10^4/s with a visibility of 96% (94%) in the H/V $(+/-)$ basis. **b.** Setup for engineering the bit-flip noise. **c.** The measurement setup. (This figure is taken from Ref. [82].)

3.5 Experimental setup

Spontaneous down conversion has been used to produce the desired four-photon state [see Fig. 3.2(a)]. With the help of polarizing beam splitters (PBSs), half-wave plates (HWPs) and conventional photon detectors, we prepare a four-qubit GHZ state, where $|0\rangle = |H\rangle$ ($|1\rangle = |V\rangle$) denotes horizontal (vertical) polarization.

We have chosen the bit-flip noise channel to demonstrate the theory introduced in this chapter. As shown in Fig. 3.2(b), the noisy quantum channels are engineered by one HWP sandwiched with two quarter-wave plates (QWPs) [87]. The HWP is switched randomly between $+\theta$ and $-\theta$ and the QWPs are set at $0°$ with respect to the vertical direction. In this way, the noise channel can be engineered with a bit-flip probability $p = \sin^2(2\theta)$. The Pauli matrix measurements required in the Bell test can be implemented by a combination of HWP, QWP and PBS [see Fig. 3.2(c)]. The fidelity of the prepared GHZ state is obtained via

3 Bell inequalities: Statistical significances in experiments

θ	p	$\mathcal{V}(\mathcal{B}_M)$	$\mathcal{E}(\mathcal{B}_M)$	$\mathcal{S}(\mathcal{B}_M)$	$\mathcal{V}(\mathcal{B}_A)$	$\mathcal{E}(\mathcal{B}_A)$	$\mathcal{S}(\mathcal{B}_A)$
$\pm 0°$	0	2.37	0.05	44.3	3.65	0.10	35.0
$\pm 2°$	0.005	2.00	0.06	33.4	3.14	0.11	29.2
$\pm 4°$	0.019	1.57	0.07	23.7	2.48	0.11	21.8
$\pm 6°$	0.043	1.13	0.07	16.2	2.05	0.11	17.8
$\pm 8°$	0.076	0.67	0.08	8.8	1.63	0.12	13.7

Table 3.1: Experimental values of the violation, the statistical error and the significance for different values of θ (and the corresponding p). $\mathcal{V}(\mathcal{B}_M), \mathcal{E}(\mathcal{B}_M), \mathcal{S}(\mathcal{B}_M)$ represent the values of $\mathcal{V}$, $\mathcal{E}$ and $\mathcal{S}$ when testing the Mermin inequality; $\mathcal{V}(\mathcal{B}_A), \mathcal{E}(\mathcal{B}_A), \mathcal{S}(\mathcal{B}_A)$ represent the corresponding values for the Ardehali inequality. Each setting $X_1X_2X_3X_4$ etc. in the Mermin inequality is measured for 800 s, while each setting $A_1X_2X_3X_4$ etc. in the Ardehali inequality is measured for 400 s. The average total count number for each inequality is about 7500.

$$F = \frac{1}{2}\text{Tr}(|0000\rangle\langle 0000| + |1111\rangle\langle 1111|) + \frac{1}{16}\langle \mathcal{B}_M \rangle. \tag{3.63}$$

Without added noise, its value is $F = 0.84 \pm 0.01$.

3.6 Experimental results

For different noise levels, the experimental results of the violation, the statistical error and the significance are shown in Table 3.1. The first observation is that, when there is no engineered noise, the violation of the Mermin inequality is smaller than the violation of the Ardehali inequality. Its significance, however, is larger than that of the Ardehali inequality; this proves that testing the Mermin inequality is a better choice to characterize the entanglement in this case. Secondly, when the noise level increases, the significance in the Mermin inequality decreases more quickly. When $\theta = \pm 6°, \pm 8°$, the significance for the Ardehali inequality is already larger than that for the Mermin inequality. Due to the experimental imperfections, the initial state to which the noise is added is not the perfect GHZ state. However, assuming an initial state like

$$\varrho(p=0) = \alpha|0000\rangle\langle 0000| + \beta|1111\rangle\langle 1111| + \gamma(|0000\rangle\langle 1111| + |1111\rangle\langle 0000|) + \frac{\lambda}{16}\mathbb{1}, \tag{3.64}$$

where $\alpha = 0.362, \beta = 0.522, \gamma = 0.398, \lambda = 0.12$ reproduces that for $p \leq 0.019$ the Mermin inequality is more significant.

3.7 Discussion

We have proved that it can be favorable to use an entanglement witness or a Bell inequality that results in a lower violation. We confirmed this experimentally using four-photon GHZ states. Our results show that the usual way of optimizing witnesses will not necessarily lead to more powerful tools for the analysis of many-particle experiments. It is important to note that when the number of photons in multi-photon experiments is increased the count rates decrease; consequently, the statistical error becomes more and more relevant.

Our results provide a direction to find powerful entanglement tests for low count rates: the observed effect relied on the fact that in the Mermin inequality only stabilizer measurements were made. There are already powerful approaches available to construct witnesses from stabilizer observables [72] and also other Mermin-like or Ardehali-like Bell inequalities have been explored [88–90]. Consequently, these approaches are promising candidates for developing sensitive analysis tools. Further, inequalities similar to witnesses have been proposed and used to characterize quantum gate fidelities [25, 91], which is another application of our theory.

4 Entanglement detection via PPT mixtures

Having studied the statistical behavior of entanglement tests in experiment, we will now pass on to the theoretical verification of genuine multipartite entanglement given that the state of the system is known without or with negligible error.[1]

We derive a general method to characterize genuine multipartite entanglement using suitable relaxations which leads to a criterion for this type of entanglement. This chapter mainly focuses on the presentation of the criterion, the idea behind it and its most important properties, while chapters 5 and 6 are then dedicated to some of its applications.

Our criterion can be considered to be a generalization of the Peres-Horodecki criterion [30, 31] to the multipartite case. We argue that the relaxed problem that we consider is good-natured, as the criterion to be introduced can be evaluated efficiently using semidefinite programming. In addition, we show that it improves existing criteria significantly and, in fact, is necessary and sufficient for permutation-invariant states of three qubits (cf. Sec. 4.4). Furthermore, we use it to estimate the volume of the set of genuinely multipartite entangled states and we apply it to the situation in which not a whole tomography, but only a few observables have been measured. Finally, in Sec. 4.5, we give an example of an analytical witness for the W state.

We note that our criterion can be applied to multi-qubit systems, continuous-variable or hybrid systems and can be evaluated, even if the mean values of only a few observables are known. Furthermore, it leads to a computable entanglement monotone for genuine multiparticle entanglement (cf. Sec. 4.3).

4.1 General idea

We start by considering a three-particle quantum state ϱ. In the following, whenever we talk about multipartite entangled states, we refer to genuine multipartite entanglement as defined in Sec. 2.1.1. We remind the reader of the fact that the general form of a biseparable state as in Eq. (2.7) cannot be easily checked for as, in principle, a search through all possible decompositions is necessary.

Thus, to characterize multipartite entanglement, we apply the method illustrated by Fig. 4.1. Instead of states like $\varrho_{A|BC}^{\text{sep}}$ that are separable with respect to a fixed bipartition, we consider a larger set of states, which can be more easily characterized. For instance, for the bipartition $A|BC$ we may

---
[1]Reprinted excerpts with permission from Ref. [92], http://prl.aps.org/abstract/PRL/v106/i19/e190502. Copyright (2011) by the American Physical Society.

4 Entanglement detection via PPT mixtures

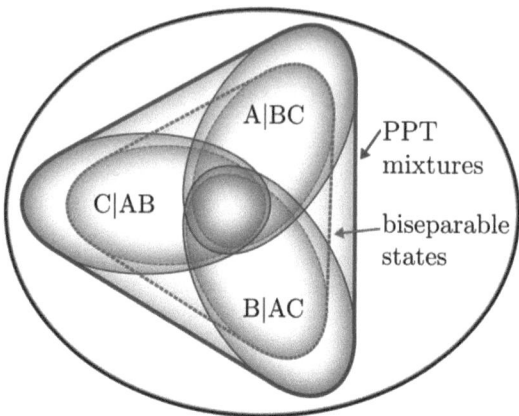

Figure 4.1: For three qubits, there are three convex sets of states that are separable with respect to a fixed bipartition, namely the bipartitions $A|BC$, $B|AC$ and $C|AB$ (thin dashed dashed borders). Their convex hull (thick dashed border) is the set of biseparable states (cf. Sec. 2.1.1). Each of the three sets is contained within the set of states that are PPT with respect to the corresponding bipartition (thin solid lines). Their convex hull forms the set of PPT mixtures (thick solid line).

consider states which have a positive partial transpose [cf. Eq. (2.9)]. According to the PPT criterion, which was given here as Theorem 7, separable states are also PPT [30, 31]. We denote such states by $\varrho_{A|BC}^{\text{ppt}}$ (and analogously for the other bipartitions).

Thus, we ask whether a state can be written as

$$\varrho^{\text{pmix}} = p_1 \varrho_{A|BC}^{\text{ppt}} + p_2 \varrho_{B|AC}^{\text{ppt}} + p_3 \varrho_{C|AB}^{\text{ppt}}. \tag{4.1}$$

Definition 26. *In the following, a state that can be written in the form of Eq. (4.1) is called a **PPT mixture**.*

Clearly, any biseparable state is a PPT mixture, so proving that a state is no PPT mixture implies genuine multipartite entanglement. There are examples of states, which are PPT with respect to any bipartition, but nevertheless multipartite entangled [93, 94]. Hence, not all multipartite entangled states can be detected in this way, but, as we will see, often the set of PPT mixtures is a very good approximation to the set of biseparable states. Finally, note that all definitions can be extended to

n particles. Also, one may use other relaxations of bipartite separability, e.g. apply the criterion of Doherty et al. [95, 96].

The advantage of considering PPT mixtures instead of biseparable states is that the set of PPT mixtures can be fully characterized with the method of linear semidefinite programming (cf. Sec. 2.6.1) — a standard problem of constrained convex optimization theory. Moreover, PPT mixtures can also be characterized analytically using the notion of entanglement witnesses which was introduced in Sec. 2.2.2.

4.2 Characterization via entanglement witnesses

We can generalize the class of decomposable entanglement witnesses of Definition 13 as follows.

Definition 27. *For more than two particles, we call a witness W **fully decomposable** if, for every subset M of all systems, it is decomposable with respect to the bipartition given by M and its complement $\overline{M}$. This means, there exist positive semidefinite operators P_M and Q_M, such that*

$$\text{for all } M : W = P_M + Q_M^{T_M}. \tag{4.2}$$

A fully decomposable witness is positive on all PPT mixtures, as it is non-negative on all states which are PPT with respect to some bipartition. But also the converse holds:

Lemma 28. *If ϱ is not a PPT mixture, then there exists a fully decomposable witness W that detects ϱ.*

Proof. The set of PPT mixtures is convex and compact. Therefore, for any state outside of it, there is a witness that is positive on all PPT mixtures. Furthermore, positivity on all states that are PPT with respect to a fixed (but arbitrary) bipartition implies that the witness is decomposable with respect to this fixed (but arbitrary) bipartition [33]. Thus, $W = P_M + Q_M^{T_M}$ for any M. □

4.3 Practical evaluation of the criterion

To find a fully decomposable witness for a given state, the convex optimization technique SDP becomes important, since it allows us to optimize over all fully decomposable witnesses. Given a multipartite state ϱ, the search is given by

$$\min \text{Tr}(W\varrho) \tag{4.3}$$
$$\text{s.t. } \text{Tr}(W) = 1 \text{ and for all } M :$$
$$W = P_M + Q_M^{T_M}, Q_M \geq 0, P_M \geq 0.$$

The free parameters are given by W and the operators P_M for every subset M. If the minimum in Eq. (4.3) is negative, ϱ is not a PPT mixture and hence genuinely multipartite entangled. The operator W for which the negative minimum is obtained is a fully decomposable witness. Note that, due to $X^{T_M} = (X^T)^{T_{\overline{M}}}$ and $X \geq 0 \Leftrightarrow X^T \geq 0$, a witness that is decomposable with respect to M is also decomposable with respect to $\overline{M}$. Thus, one needs to check only half of all subsets in practice.

Equation (4.3) has the form of a semidefinite program [cf. Eq. (2.77)]. In contrast to usual optimization problems, global optimality of an SDP can be certified and the solution can efficiently be computed via interior-point methods. In practice, Eq. (4.3) can be solved with few lines of code, using e.g. the parser YALMIP [97] and, as solvers, SeDuMi [98] or SDPT3 [99, 100]. Our implementation in MATLAB named PPTMixer can be found online [101].

Let us discuss two variations of Eq. (4.3). First, in order to reduce the number of free parameters, one can restrict oneself to witnesses W that obey $W^{T_M} \geq 0$ for all M, i.e. $P_M = 0$ for all M. In the following, we will call these witnesses **fully PPT witnesses**. For bipartite systems, decomposable witnesses and fully PPT witnesses detect the same states. For the multipartite case, fully PPT witnesses are not as good as fully decomposable witnesses, but they are easier to characterize.

Second, this SDP can also be modified to account for the case that, instead of a full tomography, only a restricted set of observables is measured. Let $\mathcal{O} = \{O_1, ... O_k\}$ be such a set of observables. Then, adding $W = \sum_{i=1}^{k} \lambda_i O_i$ to the constraints in Eq. (4.3) results in an SDP that searches for witnesses which can be evaluated knowing these observables. Note that for this program the free parameters are given by the real numbers λ_i and their number might be considerably smaller than in Eq. (4.3). If the minimum in Eq. (4.3) is non-negative, there exists a PPT mixture with expectation values $\langle O_i \rangle$. However, one may then add further observables to $\mathcal{O}$ and run the SDP again. Repeating this procedure gives more and more sensitive tests. We will discuss an example later. In practice, this program can even decide separability if the O_i have already been measured, so it can be used to gain new insights into already performed experiments.

But before proceeding to the examples, let us note three more facts. First, in the formulation no dimension of the Hilbert space is fixed. Consequently, our approach is valid for any dimension and combined with the methods of Ref. [102, 103] it can be directly used to study multipartite entanglement in continuous-variable or hybrid systems [104]. For continuous variables, it can be employed complementary to the methods of Ref. [105].

Second, our approach can also be used to *quantify* genuine multipartite entanglement. In order to do so, we modify Eq. (4.3) to obtain the following quantity:

$$N(\varrho) = - \min_{W \in \mathcal{W}} \text{Tr}(\varrho W), \tag{4.4}$$

$$\mathcal{W} = \{W | \text{ for all } M : \exists\, P_M, Q_M \text{ such that}$$
$$0 \leq P_M, Q_M \leq \mathbb{1} \text{ and } W = P_M + Q_M^{T_M}\}, \tag{4.5}$$

where M is a strict subset of the set of all qubits. Note that the class $\mathcal{W}$ consists of fully decomposable witnesses which are only normalized in a different way than before. Then, the following lemma holds.

4.3 Practical evaluation of the criterion

Lemma 29. $N(\varrho)$ *is an entanglement monotone for genuine multipartite entanglement, since it fulfills the following properties:*

(i) $N(\varrho^{\text{bs}}) = 0$ *for all biseparable states* ϱ^{bs}.

(ii) $N[\Lambda_{\text{LOCC}}(\varrho)] \leq N(\varrho)$ *for all full LOCC operations.*

(iii) $N(U_{\text{loc}} \varrho U_{\text{loc}}^\dagger) = N(\varrho)$ *for local basis changes* U_{loc}.

(iv) $N(\sum_i p_i \varrho_i) \leq \sum_i p_i N(\varrho_i)$ *holds for all convex combinations* $\sum_i p_i \varrho_i$.

Thus, $N(\varrho)$ *is a monotone for genuine multipartite entanglement. In the bipartite case, the monotone* $N(\varrho)$ *of Eq. (4.4) equals the negativity [38, 106].*

Proof. The first statement follows directly from the fact that any fully decomposable witness can only detect genuine multipartite entanglement, hence the expectation value satisfies $\text{Tr}(\varrho^{\text{bs}} W) \geq 0$ and vanishes for $W = 0$.

For the second property we effectively show $N[\Lambda(\varrho)] \leq N(\varrho)$ for all trace-preserving, completely positive operations $\Lambda(\varrho) = \sum_i K_i \varrho K_i^\dagger$ with $\sum_i K_i^\dagger K_i = \mathbb{1}$ that admit a fully separable operator-sum representation, which means that each operator $K_i = A_i \otimes B_i \otimes \cdots \otimes F_i$ has a tensor product form. This set of operations defines a superset to the set of full LOCC operations, so we verify the above property for an even larger set of possible operations [34]. Let us point out that for each operation Λ, there exists an adjoint operation $\Lambda^\dagger(Y) = \sum_i K_i^\dagger Y K_i$, that satisfies the identity $\text{Tr}[\Lambda(X)Y] = \text{Tr}[X\Lambda^\dagger(Y)]$ for all linear operators X, Y. The trace-preserving condition for Λ translates to a unital condition for the adjoint map $\Lambda^\dagger(\mathbb{1}) = \mathbb{1}$.

Let us first prove the following statement: For each trace-preserving, separable operation Λ and for any decomposable operator W the observable $\Lambda^\dagger(W)$ is decomposable as well. Suppose that $W = P + Q^{T_M}$ is an appropriate decomposition[2] with respect to a chosen partition M. Because of linearity we obtain $\Lambda^\dagger(W) = \Lambda^\dagger(P) + \Lambda^\dagger(Q^{T_M})$. First, we want to check "normalization" $0 \leq \Lambda^\dagger(P) \leq \mathbb{1}$. Complete positivity provides $\Lambda^\dagger(P) \geq 0$ since $P \geq 0$ is positive semidefinite itself. If one applies the adjoint map to $(\mathbb{1} - P) \geq 0$ and employs the unital condition one obtains

$$\Lambda^\dagger(\mathbb{1} - P) = \Lambda^\dagger(\mathbb{1}) - \Lambda^\dagger(P) = \mathbb{1} - \Lambda^\dagger(P) \geq 0, \tag{4.6}$$

hence the upper bound $\Lambda^\dagger(P) \leq \mathbb{1}$ holds as well. We can apply the same argument to $\Lambda^\dagger(Q)$ if we can fulfill the identity $\Lambda^\dagger(Q^{T_M}) = [\tilde{\Lambda}^\dagger(Q)]^{T_M}$ with $\tilde{\Lambda}$ being completely positive and unital. Using the assumed tensor product structure of each operator K_i it is straightforward to deduce the Kraus operators of the liner map $\tilde{\Lambda}$ satisfying this identity. These operators $\tilde{K}_i = \tilde{A}_i \otimes \tilde{B}_i \otimes \cdots \otimes \tilde{F}_i$ are given by $\tilde{A}_i = A_i$ if $A \notin M$ and $\tilde{A}_i = A_i^*$ if $A \in M$, and similar relations for all other operators. Let us point out that this is the only step where one explicitly needs the separable operator structure.

---
[2]The subscripts M do not matter here.

Via this statement one finally obtains

$$N[\Lambda(\varrho)] = -\min_{W \in \mathcal{W}} \text{Tr}[\Lambda(\varrho)W] = -\min_{W \in \mathcal{W}} \text{Tr}[\varrho\Lambda^\dagger(W)] \quad (4.7)$$

$$\leq -\min_{W \in \mathcal{W}} \text{Tr}(\varrho W) = N[\varrho]. \quad (4.8)$$

The inequality holds since $\Lambda^\dagger(W)$ is decomposable again, hence the optimization over the complete set of decomposable entanglement witnesses can only produce a lower (negative) expectation value.

Invariance under local basis changes is a direct consequence of the previous property. Since any local basis change U_{loc} represents an invertible full LOCC operation, one obtains

$$N(\varrho) \geq N(U_{\text{loc}}\varrho U_{\text{loc}}^\dagger) \geq N[U_{\text{loc}}^\dagger(U_{\text{loc}}\varrho U_{\text{loc}}^\dagger)U_{\text{loc}}] = N(\varrho). \quad (4.9)$$

Thus a local basis does not change the value of $N(\varrho)$.

The convexity statement

$$N\left(\sum_i p_i \varrho_i\right) = -\min_W \sum_i p_i \text{Tr}(\varrho_i W) \quad (4.10)$$

$$\leq \sum_i p_i \left[-\min_W \text{Tr}(\varrho_i W)\right] \quad (4.11)$$

$$= \sum_i p_i N(\varrho_i), \quad (4.12)$$

follows from linearity of the trace and the fact that if one performs independent optimizations then the obtained expectation value can only be smaller.

To see that our quantity equals the negativity in the bipartite case, we write

$$\text{Tr}(W\varrho) = \text{Tr}(P_A \varrho) + \text{Tr}(Q_A^{T_A} \varrho) \quad (4.13)$$

$$= \text{Tr}(P_A \varrho) + \text{Tr}(Q_A \varrho^{T_A}), \quad (4.14)$$

where we used $\text{Tr}(C^{T_A} D) = \text{Tr}(CD^{T_A})$. This expression is minimized under the constraints $\mathbb{1} \geq P_A, Q_A \geq 0$ by letting $P_A = 0$ and $Q_A = \sum_i |\phi_i\rangle\langle\phi_i|$, where $|\phi_i\rangle$ are the eigenvectors of ϱ^{T_A} that correspond to negative eigenvalues. The trace then sums over all negative eigenvalues of ϱ^{T_A}, which equals the definition of the negativity (cf. Definition 14 or Refs [38, 106]). □

Finally, as mentioned before, we remark that there are other possible choices of supersets for the set of separable states, e.g. the set of states that have a symmetric extension on a larger Hilbert space (cf. [95, 96]).

These extensions are defined as follows. For the bipartition $A|BC$, we consider states that possess a symmetric extension to k copies of system A. This means that the given state $\varrho_{A|BC}$ can be written as the reduced state of a multipartite state $\varrho_{A_1...A_k|BC}$ that is invariant under all possible permutations

4.4 Numerical examples

of the copied subsystems. Every separable state necessarily satisfies this extension condition for any number of copies and we denote states of this class by $\varrho_{A|BC}^{\mathrm{sym}_k}$. Consequently, we ask whether a three-particle state can be decomposed as

$$\varrho = p_1 \varrho_{A|BC}^{\mathrm{sym}_k} + p_2 \varrho_{B|AC}^{\mathrm{sym}_k} + p_3 \varrho_{C|AB}^{\mathrm{sym}_k}. \tag{4.15}$$

Any biseparable state can be written in this way, hence if this expansion fails, genuine multipartite entanglement is detected. This approach has appealing properties: With increasing number of copies, these supersets converge to the set of separable states [95, 96]. Moreover, it is again possible to characterize such decompositions using entanglement witnesses that can be tackled via SDP. These witnesses are such that for all possible bipartitions M, the operator W must be a bipartite entanglement witness for the case of k symmetric extensions as given in Ref. [96].

4.4 Numerical examples

Let us come back to the criterion of Eq. (4.3) and test it for some important pure three- and four-qubit states prepared in many experiments. Besides the GHZ state of n qubits as in Eq. (2.43), these are the W states of three and four qubits,

$$|W_3\rangle = (|001\rangle + |010\rangle + |100\rangle)/\sqrt{3} \tag{4.16}$$
$$|W_4\rangle = (|0001\rangle + |0010\rangle + |0100\rangle + |1000\rangle)/2 \tag{4.17}$$

and the linear cluster state of four qubits $|Cl_4\rangle$, the four-qubit Dicke state with two excitations $|D_{2,4}\rangle$ and the four qubit singlet state $|\Psi_{S,4}\rangle$, where

$$|Cl_4\rangle = (|0000\rangle + |0011\rangle + |1100\rangle - |1111\rangle)/2 \tag{4.18}$$
$$|D_{2,4}\rangle = (|0011\rangle + |1100\rangle + |0101\rangle + |0110\rangle + |1001\rangle + |1010\rangle)/\sqrt{6} \tag{4.19}$$
$$|\Psi_{S,4}\rangle = \left[|0011\rangle + |1100\rangle - \frac{1}{2}(|0101\rangle + |0110\rangle + |1001\rangle + |1010\rangle)\right]/\sqrt{3} \tag{4.20}$$

For our test, we use the white noise tolerance as a figure of merit. [3]

Definition 30. *The **white noise tolerance** of an entanglement criterion for a given state $|\psi\rangle$ is the maximal amount p_{tol} of white noise for which the state $\varrho(p_{\mathrm{tol}}) = (1 - p_{\mathrm{tol}})|\psi\rangle\langle\psi| + p_{\mathrm{tol}} \mathbb{1}/2^n$ is still detected as entangled by the criterion.*

Table 4.1 shows the white noise tolerances of our criterion, compared with the most robust criteria so far.

---
[3]This quantity is commonly used to characterize the robustness of entanglement or non locality criteria against noise [80].

state	white noise tolerances p_{tol}	
	fully decomposable	before
$\lvert GHZ_3\rangle^\star$	0.571	0.571 [16]
$\lvert GHZ_4\rangle^\star$	0.533	0.533 [16]
$\lvert W_3\rangle^\star$	0.521	0.421 [16]
$\lvert W_4\rangle$	0.526	0.444 [16]
$\lvert Cl_4\rangle^\star$	0.615	0.533 [107]
$\lvert D_{2,4}\rangle$	0.539	0.471 [108]
$\lvert \Psi_{S,4}\rangle$	0.553	0.317 [109]

Table 4.1: White noise tolerances of the fully decomposable witnesses obtained by the SDP of Eq. (4.3) compared with the corresponding tolerances of the most robust criteria known so far. For the states marked by $\star$, we verified that adding more white noise than what is tolerated by Eq. (4.3) results in a biseparable state, so the values are optimal.

Strikingly, the tolerances of the witnesses obtained by our SDP are significantly higher than previous ones. For the GHZ and the W state of three qubits and the GHZ and the linear cluster state of four qubits, we even obtain the best white noise tolerance possible, since one can show that for a larger amount of white noise the state becomes biseparable. For GHZ states, this has been shown in Ref. [16]. For the four-qubit linear cluster state, we refer to Sec. 6. For the three-qubit GHZ and W state, the fact that our criterion is necessary and sufficient even holds in the general case of a permutation-invariant state.

Lemma 31. *Any three-qubit state that is invariant under any permutation of qubits is biseparable if and only if it is a PPT mixture.*

Proof. As mentioned before, any biseparable state is a PPT mixture (due to the PPT criterion). Therefore, we need to show that, in the case of three-qubit permutation-invariant states, all PPT mixtures are biseparable. Therefore, let us start by considering such a state which can be written according to Eq. (4.1) as

$$\varrho_{\text{inv}} = p_1 \varrho_A + p_2 \varrho_B + p_3 \varrho_C \,, \qquad (4.21)$$

where the p_i are normalized and non-negative and $q_A^{T_A}$, $q_B^{T_B}$ and $q_C^{T_C}$ are positive semidefinite. Note that this notation differs slightly from Eq. (4.1), but was chosen for the sake of brevity for this proof.

Since ϱ_{inv} is invariant under permutation of qubits, one can write it as

$$\varrho_{\text{inv}} = \frac{1}{6} \sum_i \Pi_i \varrho_{\text{inv}} \Pi_i \,, \qquad (4.22)$$

where we sum over all six possible permutations of qubits. We plug Eq. (4.21) into Eq. (4.22) and

4.4 Numerical examples

explicitly write down the permutations, where P_{ij} swaps qubits i and j.

$$
\begin{aligned}
6\,(p_1\varrho_A + p_2\varrho_B + p_3\varrho_C) \\
= p_1\left(\varrho_A + P_{BC}\varrho_A P^\dagger_{BC}\right) + p_1 P_{AB}\left(\varrho_A + P_{BC}\varrho_A P^\dagger_{BC}\right) P^\dagger_{AB} \\
+ p_1 P_{AC}\left(\varrho_A + P_{BC}\varrho_A P^\dagger_{BC}\right) P^\dagger_{AC} + p_2\left(\varrho_B + P_{AC}\varrho_B P^\dagger_{AC}\right) \\
+ p_2 P_{AB}\left(\varrho_B + P_{AC}\varrho_B P^\dagger_{AC}\right) P^\dagger_{AB} + p_2 P_{BC}\left(\varrho_B + P_{AC}\varrho_B P^\dagger_{AC}\right) P^\dagger_{BC} \\
+ p_3\left(\varrho_C + P_{AB}\varrho_C P^\dagger_{AB}\right) + p_3 P_{BC}\left(\varrho_C + P_{AB}\varrho_C P^\dagger_{AB}\right) P^\dagger_{BC} \\
+ p_3 P_{AC}\left(\varrho_C + P_{AB}\varrho_C P^\dagger_{AB}\right) P^\dagger_{AC} .
\end{aligned} \quad (4.23)
$$

Now, we group the terms on the right-hand side of this equation which are PPT with respect to A. They must equal the term on the left-hand side which is PPT with respect to A, which implies that

$$
\begin{aligned}
6 p_1 \varrho_A = p_1\left(\varrho_A + P_{BC}\varrho_A P^\dagger_{BC}\right) + p_2 P_{AB}\left(\varrho_B + P_{AC}\varrho_B P^\dagger_{AC}\right) P^\dagger_{AB} \\
+ p_3 P_{AC}\left(\varrho_C + P_{AB}\varrho_C P^\dagger_{AB}\right) P^\dagger_{AC} .
\end{aligned} \quad (4.24)
$$

The same can be done for ϱ_B and ϱ_C. Thus,

$$
\begin{aligned}
6 p_2 \varrho_B = p_1 P_{AB}\left(\varrho_A + P_{BC}\varrho_A P^\dagger_{BC}\right) P^\dagger_{AB} + p_2\left(\varrho_B + P_{AC}\varrho_B P^\dagger_{AC}\right) \\
+ p_3 P_{BC}\left(\varrho_C + P_{AB}\varrho_C P^\dagger_{AB}\right) P^\dagger_{BC} .
\end{aligned} \quad (4.25)
$$

$$
\begin{aligned}
6 p_3 \varrho_C = p_1 P_{AC}\left(\varrho_A + P_{BC}\varrho_A P^\dagger_{BC}\right) P^\dagger_{AC} + p_2 P_{BC}\left(\varrho_B + P_{AC}\varrho_B P^\dagger_{AC}\right) P^\dagger_{BC} \\
+ p_3\left(\varrho_C + P_{AB}\varrho_C P^\dagger_{AB}\right) .
\end{aligned} \quad (4.26)
$$

Using these equations and the relations $P_{AB}P_{AC}P_{AB} = P_{BC}$ and $P_{AC}P_{AB}P_{AC} = P_{BC}$, one can now verify that $p_2 P_{AB}\varrho_B P^\dagger_{AB} = p_1\varrho_A$ and $p_3 P_{AC}\varrho_C P^\dagger_{AC} = p_1\varrho_A$. These equations then prove that $\varrho_A = P_{BC}\varrho_A P^\dagger_{BC}$ and therefore

$$
\varrho_A = \frac{1}{2}\left(\varrho_A + P_{BC}\varrho_A P_{BC}\right) . \quad (4.27)
$$

Now, one can write ϱ_A as a sum of two terms: One that lives in the symmetric subspace of qubits B and C and one which is separable with respect to bipartition $A|BC$. More precisely, we write ϱ_A in the basis $\{|i\rangle \otimes |\phi^{(j)}_{BC}\rangle\}$, where $i = 0, 1$ and $j = 1, 2, 3, 4$, and $|\phi^{(1)}_{BC}\rangle = |00\rangle$, $|\phi^{(2)}_{BC}\rangle = |11\rangle$, $|\phi^{(3)}_{BC}\rangle = (|01\rangle + |10\rangle)/\sqrt{2}$, $|\phi^{(4)}_{BC}\rangle = |\psi^-\rangle = (|01\rangle - |10\rangle)/\sqrt{2}$. Then, we apply P_{BC} to these basis

4 Entanglement detection via PPT mixtures

vectors, which results in

$$\varrho_A = \sum_{\substack{1 \leq j,l \leq 3 \\ i,k=0,1}} \varrho_{ij,kl} |i\rangle\langle k| \otimes |\phi_{BC}^{(j)}\rangle\langle \phi_{BC}^{(l)}| + \sum_{i,k=0,1} \varrho_{i4,k4} |i\rangle\langle k| \otimes |\psi^-\rangle\langle \psi^-| \quad (4.28)$$

$$= \sum_{\substack{1 \leq j,l \leq 3 \\ i,k=0,1}} \varrho_{ij,kl} |i\rangle\langle k| \otimes |\phi_{BC}^{(j)}\rangle\langle \phi_{BC}^{(l)}| + \tilde{\varrho}_A \otimes |\psi^-\rangle\langle \psi^-| , \quad (4.29)$$

where we defined $\tilde{\varrho}_A = \sum_{i,k=0,1} \varrho_{i4,k4} |i\rangle\langle k|$.

The two terms on the right-hand side live on different spaces. When we partially transpose the whole equation with respect to A, the two terms on the right still live on distinct spaces and must add up to the positive operator $\varrho_A^{T_A}$ on the left-hand side. Therefore, *both* terms on the right must be PPT with respect to $A|BC$. As the first term is PPT and lives on a space of dimension $2 \otimes 3$, it must be separable [31]. The second term is clearly separable. Therefore, ϱ_A is separable (with respect to $A|BC$).

The same line of argument also shows that ϱ_B and ϱ_C are separable and therefore ϱ_{inv} is biseparable. □

This shows that our criterion is indeed optimal for the states in Table 4.1 marked by *.

To show that the criterion of Eq. (4.3) works well for higher-dimensional states and a restricted set of observables, we consider the four-qubit Dicke state with two excitations $|D_{2,4}\rangle$ [cf. Eq. (4.19)]. For this state, the SDP yields the witness

$$W_D = \frac{1}{16} \left[\mathbb{1} + \alpha_1 X^{\otimes 4} + \alpha_1 Y^{\otimes 4} + \alpha_2 Z^{\otimes 4} + \alpha_3 \left(X_1 X_2 Y_3 Y_4 + \text{perms} \right) \right.$$
$$+ \alpha_4 \left(Z_1 Z_2 Y_3 Y_4 + \text{perms} \right) + \alpha_4 \left(Z_1 Z_2 X_3 X_4 + \text{perms} \right)$$
$$+ \alpha_5 \left(X_1 X_2 \mathbb{1}_3 \mathbb{1}_4 + \text{perms} \right) + \alpha_5 \left(Y_1 Y_2 \mathbb{1}_3 \mathbb{1}_4 + \text{perms} \right)$$
$$\left. + \alpha_6 \left(Z_1 Z_2 \mathbb{1}_3 \mathbb{1}_4 + \text{perms} \right) \right] . \quad (4.30)$$

Here, $X_1 X_2 Y_3 Y_4 + \text{perms}$ is the sum over all distinct permutations of $X_1 X_2 Y_3 Y_4$. Moreover, $\alpha_1 = 0.014, \alpha_2 = -0.095, \alpha_3 = 0.0046, \alpha_4 = 0.16, \alpha_5 = -0.14, \alpha_6 = -0.15$.

W_D only consists of $\mathcal{O} = \{X^{\otimes 4}, Y^{\otimes 4}, Z^{\otimes 4}, X_1 X_2 Y_3 Y_4, X_1 X_2 Z_3 Z_4, Y_1 Y_2 Z_3 Z_4\}$, distinct permutations of these observables and other observables that can be measured in the same run. For example, a local measurement of $X_1 X_2 X_3 X_4$ yields knowledge of the expectation value of $X_1 X_2 \mathbb{1}_3 \mathbb{1}_4$. The SDP finds a witness consisting of $O_1 = X^{\otimes 4}, O_2 = Y^{\otimes 4}$ and observables obtained by replacing some Paulis by the identity. Already with these observables, the white noise tolerance is $p_{\text{tol}}^{(2)} \approx 0.29495$. We can proceed in this way and use additional observables O_i from the set $\mathcal{O}$ — including their permutations and observables obtained by replacing Pauli operators by $\mathbb{1}$ — to produce strictly stronger witnesses $W_D^{(i)}$. Their white noise tolerances $p_{\text{tol}}^{(i)}$ are $p_{\text{tol}}^{(3)} \approx 0.38379, p_{\text{tol}}^{(4)} \approx 0.38383, p_{\text{tol}}^{(5)} \approx 0.45200$ and finally $p_{\text{tol}}^{(6)} \approx 0.53914$ as in Table 4.1, since $W_D = W_D^{(6)}$.

4.5 An analytical witness for the W state

Finally, we compute a lower bound on the volume of genuinely multipartite entangled states. We created samples of 10^4 random mixed three-qubit states uniformly distributed with respect to the Hilbert-Schmidt distance (or the Bures distance) and check whether they are genuinely multipartite entangled. 6.28 % (Bures: 10.32 %) were detected by fully decomposable and 0.44 % (Bures: 1.06 %) by fully PPT witnesses. As expected, fully PPT witnesses detect fewer states.

While the problem can still be tackled numerically for six or seven qubits, in recent experiments up to 14 ions have been coherently manipulated [24]. Therefore, an analytical treatment is desirable to obtain witnesses for an arbitrary number of qubits. Before we do so in Sec. 5, let us first examine the state $|W_3\rangle$ for which the PPT witness produced by our SDP can also be understood analytically.

4.5 An analytical witness for the W state

Let us consider the W state of three qubits $|W_3\rangle$ from Eq. (4.16). For this state, the best PPT witness that the program of Eq. (4.3) finds (with $P_M = 0$ for all M) is given by

$$W_{W_3} = 0.253\,(|001\rangle\langle001| + |010\rangle\langle010| + |100\rangle\langle100|) - 0.380|W_3\rangle\langle W_3|$$
$$+ 0.310|000\rangle\langle000| + 0.103\,(|011\rangle\langle011| + |101\rangle\langle101| + |110\rangle\langle110|)\,. \tag{4.31}$$

Since these witness has been obtained numerically, we do not have an analytical expression for the occurring coefficients. However, the following lemma allows us to understand these values analytically.

Lemma 32. *Among all witnesses that detect the W_3 state and have the form*

$$W = \alpha\,(|001\rangle\langle001| + |010\rangle\langle010| + |100\rangle\langle100|) + \beta|W_3\rangle\langle W_3|$$
$$+ \gamma|000\rangle\langle000| + \delta\,(|011\rangle\langle011| + |101\rangle\langle101| + |110\rangle\langle110|) + \epsilon|111\rangle\langle111| \tag{4.32}$$

with $\alpha, \beta, \gamma, \delta, \epsilon \in \mathbb{R}$, the witness

$$W_{W_3} = \frac{2}{15}\left(-3 + 2\sqrt{6}\right)(|001\rangle\langle001| + |010\rangle\langle010| + |100\rangle\langle100|) + \frac{1}{5}\left(3 - 2\sqrt{6}\right)|W_3\rangle\langle W_3|$$
$$+ \frac{1}{5}\left(4 - \sqrt{6}\right)|000\rangle\langle000| + \frac{1}{15}\left(4 - \sqrt{6}\right)(|011\rangle\langle011| + |101\rangle\langle101| + |110\rangle\langle110|)\,. \tag{4.33}$$

has the highest white noise tolerance.

Proof. Since Eq. (4.32) is invariant under permutation of qubits, it is enough to consider the partial transpose with respect to A. Note that positivity of W^{T_A} implies positivity of $W_{T_{BC}}$. The eigenvalues of the witness W in Eq. (4.32) are given by

$$\lambda_1 = \alpha, \ \lambda_2 = \alpha + \frac{1}{3}\beta, \ \lambda_3 = \alpha + \frac{2}{3}\beta$$
$$\lambda_4 = \delta, \ \lambda_{5,6} = \frac{1}{6}\left[3\gamma + 3\delta \pm \sqrt{8\beta^2 + 9(\gamma - \delta)^2}\right]. \tag{4.34}$$

As the white noise tolerance is given by Eq. (5.79), we need to minimize $\text{Tr}(W)$ and to maximize the absolute value of $\langle W_3|W|W_3\rangle$ while keeping W^{T_A} positive semidefinite. We have $\text{Tr}(W) = 3\alpha + \beta + \gamma + 3\delta + \epsilon$ and $\langle W_3|W|W_3\rangle = \alpha + \beta$.

First, we note that we can set β to any value, since we are allowed to multiply W by a factor which does not change the white noise tolerance [cf. Eq. (5.79)]. Since $\langle W_3|W|W_3\rangle = \alpha + \beta$ must be negative while $\lambda_1 = \alpha$ cannot be negative, we set $\beta = -1$. Then, the minimal possible value for α is $\alpha = \frac{2}{3}$. Equation (4.34) shows that the smallest allowed value we can choose for ϵ in order to minimize $\text{Tr}(W)$ is $\epsilon = 0$.

The smallest possible values of γ and δ can be found as follows: Since

$$\lambda_6 = \frac{1}{6}\left[3\gamma + 3\delta - \sqrt{8\beta^2 + 9(\gamma - \delta)^2}\right] \leq \lambda_5, \tag{4.35}$$

we do not need to consider λ_5 when searching for these values. Since λ_4 cannot be negative, δ must be non-negative. Also, γ cannot be negative since otherwise the operator in Eq. (4.32) would not be a witness. Thus, a short calculation shows that $\lambda_6 \geq 0$ is equivalent with $\gamma\delta \geq \frac{2}{9}$. The linear function $\text{Tr}(W) = 1 + \gamma + 3\delta$ takes its minimum on the boundary of the allowed area, i.e. for $\gamma\delta = \frac{2}{9}$ and $\delta \geq 0$. It is now easy to see (using e.g. a Lagrange multiplier) that $\text{Tr}(W)$ is minimized for $\gamma = 3\delta = \sqrt{\frac{2}{3}}$. Altogether, we thus have

$$\alpha = \frac{2}{3}, \ \beta = -1, \ \gamma = 3\delta = \sqrt{\frac{2}{3}}, \ \epsilon = 0. \tag{4.36}$$

Up to a normalization constant, this coincides with Eq. (4.34). □

Note that, in Sec. 5, we will present various different analytical construction methods for graph state witnesses.

4.6 Discussion

In this chapter, we presented an easily implementable criterion for genuine multipartite entanglement. We demonstrated its high robustness, showed that it is necessary and sufficient for permutation-invariant three-qubit states and we connected it to entanglement measures. Moreover, we presented witnesses for the Dicke state of four qubits with two excitations and the W state of three qubits.

4.6 Discussion

Due to its versatility, the presented criterion can be used to characterize the entanglement in various physical systems, e.g. in ground states of spin models undergoing a quantum phase transition. Moreover, we believe that, as an easy-to-use scheme, it will be valuable for the analysis of experimental data that do not constitute a whole tomography. In order to extend the criterion to a larger number of particles, it would be interesting to find classes of states to which it can be applied analytically (cf. Sec. 5) and classes for which the presented criterion is necessary and sufficient (cf. Sec. 6).

5 Entanglement witnesses for graph states

The last chapter was dedicated to presenting a new criterion for genuine multipartite entanglement, illustrating its general idea and properties, deriving a monotone and showing its performance on selected states. The criterion was applied to some important states as the W and the GHZ state for three and four qubits, and, for four qubits, the linear cluster state, the singlet and the Dicke state of two excitations. Its white noise tolerance turned out to be higher than in previous criteria. Moreover, the case in which no fully tomography, but only a restricted set of observables has been measured, was considered.

In this chapter, we turn to an analytical characterization. This analytical treatment enables us to apply the criterion and the notion of fully decomposable witnesses to states of an arbitrary number of qubits and is therefore an important step to allow for its application in experiments involving a higher number of qubits.[1]

We use our approach of the last chapter to develop a general theory of witnesses for graph states (cf. Sec. 2.5). The main results of this chapter can be grouped into two parts: First, we provide entanglement criteria, so-called entanglement witnesses (cf. Sec. 2.2.2), for all graph states up to six qubits. These witnesses are optimal in the framework of the last chapter, they detect more states than the graph state witnesses known so far and thus require a lower fidelity when measured in an experiment.

Second, we extend our results to arbitrary qubit numbers by providing a general theory of how to construct witnesses for arbitrary graph states. In many cases, these witnesses improve the best known witnesses so far and have white noise tolerances that approach one for an increasing particle number. This implies that for this type of noise the state fidelity can decrease exponentially with the number of qubits, but still entanglement is present and can be detected. Moreover, this improvement comes with very low experimental costs, since it is realized by measuring one additional setting in the experiment. Furthermore, a similar improvement can be achieved for witnesses that require only two settings to be measured [72], which results in improved witnesses that consist of only two experimental settings in total.

This chapter is structured as follows. In Sec. 5.1, we show that the criterion of Eq. (4.3) can be reduced from a general semidefinite program (cf. Definition 17) to a linear program (cf. Definition 18) in the case of graph-diagonal states.

Then, we will first consider the class of fully decomposable witnesses given in Definition 27. This is done in Sec. 5.2. We provide such entanglement witnesses for all graph states of up to six qubits

---

[1]Reprinted excerpts with permission from Ref. [75], http://pra.aps.org/abstract/PRA/v84/i3/e032310. Copyright (2011) by the American Physical Society.

in Sec. 5.2.1. Then, in Sec. 5.2.2, we present analytical construction methods. We provide examples and give an extended construction for particular states in Sec. 5.2.3 including further examples.

In Sec. 5.3, we move on to another class of witnesses, the fully PPT witnesses which are easier to characterize and which were introduced in Sec. 4.2. Here, we do not only provide a construction method for witnesses of this class (in Sec. 5.3.1), but we can extend it to an even larger number of graph states compared with the case of fully decomposable witnesses. We present this extension in Sec. 5.3.2. In order to illustrate that the presented methods can be exploited further, we provide a witness for the 2D cluster state (Sec. 5.3.3).

Finally, we discuss the entanglement monotone of Eq. (4.4) for genuine multiparticle entanglement and show that graph states are the maximally entangled states for this entanglement measure. In the conclusion, we discuss our results and possible extensions for the future. In order to make this chapter as readable as possible, we grouped nearly all proofs in the last section of this chapter.

5.1 Graph-diagonal states

Let us first consider the form of the criterion in Eq. (4.3) for graph-diagonal states. Note that any state can be transformed into a graph-diagonal state by local transformations [76]. Since local operations cannot create entanglement, the presence of entanglement in the transformed, graph-diagonal state indicates that the original state was entangled.

Here, we will see that for graph-diagonal states, the corresponding search for an optimal fully decomposable entanglement witness can w.l.o.g. be restricted to graph-diagonal witnesses, for which also the operators P_M and Q_M are graph-diagonal. This is summarized in the following lemma.

Lemma 33. *For any graph diagonal state $\varrho_G = \sum_{\vec{k}} s_{\vec{k}} |\vec{k}\rangle\langle\vec{k}|$, the search for an optimal fully decomposable entanglement witness given by Eq. (4.3), can w.l.o.g. be restricted to a graph-diagonal form, i.e., to a linear program given by*

$$\min \operatorname{Tr}(W_G \varrho_G) \tag{5.1}$$

$$\text{s.t. } W_G = \sum w_{\vec{k}} |\vec{k}\rangle\langle\vec{k}|, \operatorname{Tr}(W_G) = 1 \text{ and for all } M:$$

$$W_G = P_M + Q_M^{T_M}, P_M \geq 0, Q_M \geq 0, \tag{5.2}$$

$$P_M = \sum p_{\vec{k}}^M |\vec{k}\rangle\langle\vec{k}|, Q_M = \sum q_{\vec{k}}^M |\vec{k}\rangle\langle\vec{k}|.$$

The proof is given in Sec. 5.6.

This lemma has the following important implications: First, the optimization problems simplifies to a linear program, which are in general easier to solve than general semidefinite programs. Second, it provides a great simplification in order to derive analytic witnesses, because we know that there is an optimal witness which is diagonal in the graph state basis. Also, checking positivity of any operator simplifies to verifying non-negativity within the graph state basis. Instead of testing positivity of a whole matrix, it is enough to consider products of generators g_i and sums thereof [cf. Eq. (2.69)].

Third, let us point out that this lemma also implies that, if a state is a PPT mixture, each PPT state in its decomposition can be assumed to be graph-diagonal as well. Finally, note that a similar statement as Lemma 33 holds for PPT witnesses as well.

5.2 Fully decomposable witnesses

In this section, we present a general theory for fully decomposable witnesses of graph states. First, in Sec. 5.2.1, we provide fully decomposable witnesses for all LU-equivalence classes of graph states up to six qubits. These witnesses are obtained by the criterion of Eq. (4.3). The graph states are given in Fig. 2.2, while the witnesses' white noise tolerances are given in Table 5.1. The witnesses can be found in Sec. 5.7.

Moreover, we introduce an analytical construction method for fully decomposable witnesses of general graph states in Sec. 5.2.2. This is this section's main result and is formulated in Lemma 34.

We provide specific examples in Sec. 5.2.2. Finally, we show how to construct witnesses that detect even more states using the witnesses of Lemma 34. This result is given as Lemma 36 in Sec. 5.2.3. Again, we give examples in Sec. 5.2.3.

5.2.1 Graph states up to 6 qubits

We now apply the criterion of Eq. (4.3) to certain graph states. To this end, we implemented it as a semidefinite program using the parser YALMIP [97] in combination with the solver modules SeDuMi [98] or SDPT3 [99, 100] in MATLAB. The program we wrote is called PPTMixer and can be found online [101].

As mentioned before, there are 19 LU-equivalence classes of connected graph states up to six qubits. We apply our criterion to one state of each class (cf. Fig. 2.2), obtaining the witnesses given in Sec. 5.7. By applying the rules presented in Sec. 2.5.3 and in Ref. [77], it is possible to transform these into witnesses for any graph state of up to six qubits.

Let us have a closer look at the witnesses of Sec. 5.7. A widely-used indicator for how robust a witness is to noise in an experiment is the white noise tolerance given in Definition 30. Note that the criterion of Eq. (4.3) provides witnesses with the highest possible white noise tolerance among all fully decomposable witnesses. This can be seen by noting that both $\mathrm{Tr}(W|G\rangle\langle G|)$ and $\mathrm{Tr}[W\varrho(p_{\mathrm{tol}})]$ reach their minimum for the same normalized witness W, since $\mathrm{Tr}(Wp_{\mathrm{tol}}\mathbb{1}/2^n) = p_{\mathrm{tol}}/2^n$ is independent of W. Thus, the witness that one obtains for the state $|G\rangle$ is also a witness for $\varrho(p_{\mathrm{tol}})$. In Table 5.1, we give the white noise tolerances of these witnesses.

Now, let us present some of these witnesses as examples. Note that the SDP yields witnesses whose trace is normalized to one. In order to make the structure of the witnesses more evident, we renormalized them for each state $|G\rangle$, such that $\langle G|W|G\rangle = -1/2$.

For the GHZ states of three to six qubits (cf. states No. 2, No. 3, No. 5 and No. 9 in Fig. 2.2), we obtain the well-known projector witnesses $W_{\mathrm{proj}} = \frac{1}{2}\mathbb{1} - |G\rangle\langle G|$. Since it is known that $(1-p)|GHZ_n\rangle\langle GHZ_n| +$

63

5 Entanglement witnesses for graph states

is biseparable for $p \geq [2(1 - 2^{-n})]^{-1}$ [16], these witnesses have the maximal possible white noise tolerance.

The linear cluster state of four qubits $|Cl_4\rangle$, labelled as state No. 4 in Fig. 2.2, is detected by the witness

$$W_{Cl4} = \frac{1}{2} - |Cl_4\rangle\langle Cl_4| - \frac{1}{2}\gamma_1^-\gamma_4^-, \tag{5.3}$$

where we defined $\gamma_i^{\pm} = (\mathbb{1} \pm g_i)/2$, g_i being the generators of the stabilizer group of $|Cl_4\rangle$, for the sake of a compact notation. Note that, alternatively, one can write $\gamma_1^-\gamma_4^- = (\mathbb{1} - g_1)/2(\mathbb{1} - g_4)/2 = \sum_{i,j \in \{0,1\}} |1ij1\rangle\langle 1ij1|$ in the graph state basis. We will gain a deeper understanding of the structure of this witness in the next section.

Let us use the witness of Eq. (5.3) as an example of how the witnesses are transformed when the graph is transformed by local complementations, which corresponds to an application of local unitaries to the graph state (cf. Sec. 2.5.3). The four-qubit linear cluster state can be transformed into the four-qubit ring cluster state as shown in Fig. 5.1 via local complementations on the qubits 2, 3 and 2 again. We denote the graph that is obtained after the first step by $\tilde{G}$, the graph after the second step by $\bar{G}$ and the final ring graph by G'. Their generators according to Eq. (2.57) are denoted by $\tilde{g}_i$, $\bar{g}_i$ and g'_i, respectively. Application of the transformation rules for the generators in Eq. (2.75) results in the transformations

$$W_{Cl4} = \frac{1}{2} - |Cl_4\rangle\langle Cl_4| - \frac{1}{8}(g_1 - \mathbb{1})(g_4 - \mathbb{1}) \tag{5.4}$$

$$\xrightarrow{LC_2} \frac{1}{2} - |Cl_4\rangle\langle Cl_4| - \frac{1}{8}(\tilde{g}_1\tilde{g}_2 - \mathbb{1})(\tilde{g}_4 - \mathbb{1}) \tag{5.5}$$

$$\xrightarrow{LC_3} \frac{1}{2} - |Cl_4\rangle\langle Cl_4| - \frac{1}{8}(\bar{g}_1\bar{g}_3\bar{g}_2\bar{g}_3 - \mathbb{1})(\bar{g}_3\bar{g}_4 - \mathbb{1}) \tag{5.6}$$

$$= \frac{1}{2} - |Cl_4\rangle\langle Cl_4| - \frac{1}{8}(\bar{g}_1\bar{g}_2 - \mathbb{1})(\bar{g}_3\bar{g}_4 - \mathbb{1}) \tag{5.7}$$

$$\xrightarrow{LC_1} \frac{1}{2} - |Cl_4\rangle\langle Cl_4| - \frac{1}{8}(g'_1g'_2 - \mathbb{1})(g'_4g'_1g'_3g'_1 - \mathbb{1}) \tag{5.8}$$

$$= \frac{1}{2} - |Cl_4\rangle\langle Cl_4| - \frac{1}{8}(g'_1g'_2 - \mathbb{1})(g'_3g'_4 - \mathbb{1}) \tag{5.9}$$

$$\tag{5.10}$$

Strikingly, the similar state No. 6, which we call Y_5 state, is detected by a similar witness which has, however, some additional terms. The witness is given by

$$W_{G6} = \frac{1}{2} - |G\rangle\langle G| - \frac{1}{2}\gamma_1^-\gamma_4^- - \frac{1}{2}\gamma_1^+\gamma_4^-\gamma_5^-. \tag{5.11}$$

For the symmetrized version of this state, state No. 11 (or H_6 state), we obtain a witness with even

5.2 Fully decomposable witnesses

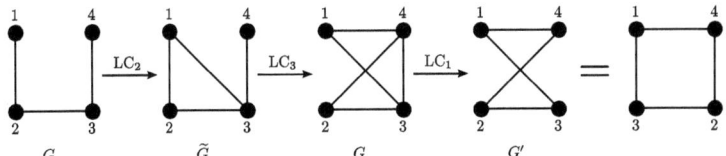

Figure 5.1: The linear four-qubit cluster state is LU-equivalent with the four-qubit ring cluster state as they can be transformed into each other by local complementations (LC).

more terms, namely

$$W_{G11} = \frac{1}{2}\mathbb{1} - |G\rangle\langle G| - \frac{1}{2}\gamma_1^- \gamma_4^- - \frac{1}{2}\gamma_1^+ \gamma_2^- \gamma_4^- - \frac{1}{2}\gamma_1^- \gamma_3^- \gamma_4^+ - \frac{1}{2}\gamma_1^+ \gamma_2^- \gamma_3^- \gamma_4^+ . \quad (5.12)$$

The special structure of these witnesses motivates an analytical investigation. In fact, we will gain more insight on the witness W_{G6} and W_{G11} in Section 5.2.3.

5.2.2 Analytical construction methods

In this section, we present an analytical method to construct fully decomposable witnesses for arbitrary graph states. This construction method results in witnesses which are a generalization of the linear cluster state witnesses in Eq. (5.3). First, we have a closer look at these witnesses for linear cluster states, before we then generalize it to arbitrary graph states in Lemma 34 in the second half of this section.

Linear cluster state

We have pointed out that the witness W_{Cl4} of Eq. (5.3) is a witness for the four-qubit linear cluster state. For the seven-qubit linear cluster state $|Cl_7\rangle$ shown in Fig. 5.2 a), there exists a similar witness

$$W_{Cl7} = \frac{1}{2}\mathbb{1} - |Cl_7\rangle\langle Cl_7| - \frac{1}{2}\left(\gamma_1^- \gamma_4^- \gamma_7^- + \gamma_1^+ \gamma_4^- \gamma_7^- + \gamma_1^- \gamma_4^+ \gamma_7^- + \gamma_1^- \gamma_4^- \gamma_7^+\right). \quad (5.13)$$

W_{Cl7} is a fully decomposable witness. However, since W_{Cl7} was not obtained from our SDP, but via Lemma 34, there are — most likely — fully decomposable witnesses for $|Cl_7\rangle$ with a higher white noise tolerance. This is in contrast to W_{Cl4} which was obtained by the semidefinite program and therefore has the maximal white noise tolerance among the fully decomposable witnesses.

W_{Cl7} has a very particular structure. The qubits i whose generators g_i appear in the witness are indicated by circles in Fig. 5.2 a). Let us denote the set of these qubits by $\mathcal{B}$. One can see that each two qubits in $\mathcal{B}$ have at least two other qubits between them. Moreover, the terms $\gamma_1^\pm \gamma_4^\pm \gamma_7^\pm$ in Eq. (5.13) all contain two or more minus signs. It turns out that witnesses of this kind can be

5 Entanglement witnesses for graph states

state	white noise tolerance
No. 1, Bell state	$p_{\text{tol}} = \frac{2}{3} \approx 0.667$
No. 2, GHZ_3	$p_{\text{tol}} = \frac{4}{7} \approx 0.571$
No. 3, GHZ_4	$p_{\text{tol}} = \frac{8}{15} \approx 0.533$
No. 4, Cl_4	$p_{\text{tol}} = \frac{8}{13} \approx 0.615$
No. 5, GHZ_5	$p_{\text{tol}} = \frac{16}{31} \approx 0.516$
No. 6, Y_5	$p_{\text{tol}} = \frac{16}{25} = 0.64$
No. 7, Cl_5	$p_{\text{tol}} = \frac{16}{25} = 0.64$
No. 8, R_5	$p_{\text{tol}} = \frac{12}{19} \approx 0.632$
No. 9, GHZ_6	$p_{\text{tol}} = \frac{32}{63} \approx 0.508$
No. 10	$p_{\text{tol}} = \frac{32}{49} \approx 0.653$
No. 11, H_6	$p_{\text{tol}} = \frac{32}{45} \approx 0.711$
No. 12, Y_6	$p_{\text{tol}} = \frac{32}{45} \approx 0.711$
No. 13, E_6	$p_{\text{tol}} = \frac{32}{45} \approx 0.711$
No. 14, Cl_6	$p_{\text{tol}} = \frac{128}{179} \approx 0.715$
No. 15	$p_{\text{tol}} = \frac{32}{47} \approx 0.681$
No. 16	$p_{\text{tol}} = \frac{8}{11} \approx 0.727$
No. 17	$p_{\text{tol}} \approx 0.696$
No. 18, R_6	$p_{\text{tol}} \approx 0.667$
No. 19	$p_{\text{tol}} = \frac{2}{3} \approx 0.667$

Table 5.1: For graph states of up to six qubits, there are 19 classes of states which are equivalent under LU operations. Here, we show one state of each class. Using the presented criterion, one obtains a witness for each of these states (cf. Sec. 5.7) which have the white noise tolerances given here.

constructed for general graph states.

Arbitrary graph states

The construction in Eq. (5.13) can be generalized in the following way:

5.2 Fully decomposable witnesses

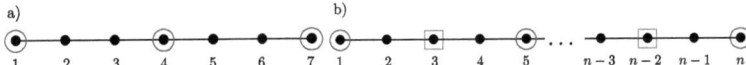

Figure 5.2: For the linear cluster state, we construct several witnesses. In a), the qubits in $\mathcal{B}$ (marked by circles) can be used to construct the fully decomposable witness of Eq. (5.13) using Lemma 34. b) illustrates the construction method of Lemma 38 which yields a fully PPT witness. Qubits in $\mathcal{B}_1$ are marked by circles, qubits in $\mathcal{B}_2$ by squares.

Lemma 34. *Given a connected graph state $|G\rangle$. Let $\mathcal{B} = \{\beta_i\}$ be a subset of the set of all qubits such that any two qubits in $\mathcal{B}$ are neither neighbors of each other nor have a neighbor in common. We define $b = |\mathcal{B}|$. Let $\sum_{\vec{s}}$ be the sum over all vectors $\vec{s}$ of length b with elements $s_i = \pm 1$ that contain at least two elements which equal -1, i.e., $\sum_{i=1}^{b} s_i \leq b - 4$. In this case,*

$$W_G = \frac{1}{2}\mathbb{1} - |G\rangle\langle G| - \frac{1}{2}\sum_{\vec{s}} \prod_{i \in \mathcal{B}} \gamma_i^{s_i} \qquad (5.14)$$

is a fully decomposable witness for $|G\rangle$.

For the detailed proof, we refer to Sec. 5.6. Its main idea is to construct a suitable positive operator P_M for every subset M, such that $(W - P_M)^{T_M} = Q_M$ is positive semidefinite.

Furthermore, the proof takes advantage of the fact that, besides $|G\rangle\langle G|$, all terms in Eq. (5.14) are invariant under any partial transposition T_M, since the identity is diagonal in any basis and there are no two generators g_i in the product that are neighbors of each other. However, products of non-neighboring generators are only tensor products of the Pauli matrices X, Z and the identity all of which are invariant under transposition. Moreover, the proof is simplified by W_G being diagonal in the graph state basis.

Note that in many cases, the choice of subset $\mathcal{B}$ is not unique. For the seven-qubit linear cluster state, instead of the choice $\mathcal{B} = \{1, 4, 7\}$ which results in the witness of Eq. (5.13), the choices $\mathcal{B} = \{1, 6\}$ or $\mathcal{B} = \{2, 5\}$ would also be valid. However, these sets would lead to witnesses that have a lower white noise tolerance.

It turns out that for many graph states, the white noise tolerances of witnesses constructed according to Lemma 34 converge to one for an increasing particle number. More precisely, this is the case for graph states that can be defined for an arbitrary number of qubits such that, when increasing the number of qubits, also the number of qubits in $\mathcal{B}$ grows. This includes the linear cluster state, the 2D cluster state for n qubits and the ring cluster state. It does not include GHZ states, since for any number of qubits, no set $\mathcal{B}$ (of more than one qubit) that contains only qubits with non-overlapping neighborhoods can be defined on the GHZ state. Let us formulate this observation as a corollary:

Corollary 35. *Let $|G_n\rangle$ be a graph state of n qubits and $\mathcal{B}(n)$ a subset of these n qubits with the properties as in Lemma 34. Let W_{G_n} be a witness for $|G_n\rangle$ as in Eq. (5.14). Then, the white noise* tolerance of W_{Gn} with respect to $|G_n\rangle$ is given by

$$p(n) = \left(1 - 2^{-n+1} + 2^{-|\mathcal{B}(n)|}(|\mathcal{B}(n)| + 1)\right)^{-1}. \tag{5.15}$$

For a family of graph states on any number of qubits n with $|\mathcal{B}(n)| \xrightarrow{n \to \infty} \infty$, this expression implies

$$p(n) \xrightarrow{n \to \infty} 1. \tag{5.16}$$

For high particle numbers, the fidelity F_{req} required to detect the state $\varrho = (1-p)|G_n\rangle\langle G_n| + p\mathbb{1}/2^n$ is given by $F_{\text{req}} \approx |\mathcal{B}(n)|2^{-|\mathcal{B}(n)|}$ and therefore vanishes exponentially fast.

For the proof, we refer to Sec. 5.6.

Note that this behavior of the white noise tolerance has been found in Ref. [92]. Moreover, entanglement criteria for Dicke states that also exhibit a white noise tolerance which converges to one have been found recently in Ref. [108].

Examples

2D cluster state — Let us consider a 2D cluster state of 16 qubits as given in Fig. 5.3 a). To construct a witness according to Lemma 34, one could choose $\mathcal{B} = \{1, 4, 10, 16\}$ as indicated by circles. However, it would also be possible to choose qubit 13 instead of qubit 10. In both cases, the white noise tolerance is $p_{\text{tol}} = (1 - 2^{-15} + 5 \cdot 2^{-4})^{-1} \approx 0.762$.

Other graph states — Consider state No. 13, the E_6 state, of Fig. 2.2. Here, $\mathcal{B} = \{1, 5, 6\}$ would be a valid choice.

For state No. 11, the H_6 state, $\mathcal{B} = \{1, 4\}$ is a possible choice. However, one could have also selected $\mathcal{B} = \{1, 3\}$, $\mathcal{B} = \{2, 3\}$ or $\mathcal{B} = \{2, 4\}$. Indeed, in the next section, we will see that all these choices can be combined to construct an even better witness, namely the witness of Sec. 5.7 which is obtained by our SDP. As mentioned before, the corresponding white noise tolerances are given in Table 5.1.

5.2.3 Extended construction method

Although Lemma 34 can be applied to many graph states, for most graph states there exist witnesses with a higher white noise tolerance (cf. Sec. 5.7). In this section, we provide an extended construction method that, for some states, allows one to subtract additional terms from the witnesses constructed by Lemma 34. This extended method can be applied to, e.g., the states No. 6 (Y_5) and No. 11 (H_6) of Fig. 2.2 to obtain the witnesses of Eqs. (5.11) and (5.12).

Lemma 36. *Given a connected graph state $|G\rangle$ and m subsets $\mathcal{B}_i$ of its qubits that fulfill the following two conditions:*

5.2 Fully decomposable witnesses

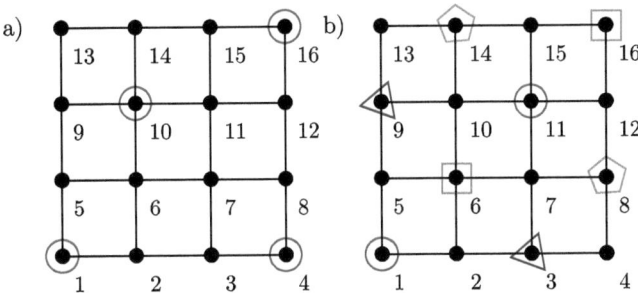

Figure 5.3: We illustrate two different ways to construct witnesses for the 2D cluster state. In a), the circles mark qubits that belong to $\mathcal{B}$, which can be used to construct a fully decomposable witness according to Lemma 34 (or a fully PPT witness using Lemma 37). In b), we illustrate the method of Lemma 38 which results in a fully PPT witness. For this, one needs to define the sets $\mathcal{B}_1$ (indicated by circles), $\mathcal{B}_2$ (triangles), $\mathcal{B}_3$ (squares) and $\mathcal{B}_4$ (pentagons).

(i) *No two qubits in a set $\mathcal{B}_i$ have a neighbor in common or are neighbors of each other.*

(ii) *Any two qubits $\beta_j^{(i)} \in \mathcal{B}_i$ and $\beta_l^{(k)} \in \mathcal{B}_k$ from two different subsets either have the same neighborhood or no common neighbor at all.*

Moreover, let W_i be the fully decomposable witnesses that one can construct from the subsets $\mathcal{B}_i$ according to Lemma 34. Then,

$$W = \sum_{\vec{k} \in \{0,1\}^n} |\vec{k}\rangle\langle\vec{k}| \min_{i=1,\ldots,m} \langle\vec{k}|W_i|\vec{k}\rangle \qquad (5.17)$$

is a fully decomposable witness. Note that W is clearly better than any of the witnesses W_i alone.

Note that it is possible, according to conditions (i) and (ii), that a qubit is in more than one subset $\mathcal{B}_i$. The proof of Lemma 36 is given in Sec. 5.6. Let us present some examples of the witnesses constructed in this lemma.

Examples

State No. 6 (Y_5) — Consider state No. 6 of Fig. 2.2. Here, the subsets $\mathcal{B}_i$ are given by $\mathcal{B}_1 = \{1,4\}$ and $\mathcal{B}_2 = \{5,4\}$, which fulfill the conditions of Lemma 36, since the neighborhoods of qubits 1 and

5 coincide, $\mathcal{N}(1) = \mathcal{N}(5)$. Lemma 34 then yields the two witnesses

$$W_1 = \frac{1}{2} - |G\rangle\langle G| - \frac{1}{2}\gamma_1^-\gamma_4^-, \qquad (5.18)$$

$$W_2 = \frac{1}{2} - |G\rangle\langle G| - \frac{1}{2}\gamma_5^-\gamma_4^-. \qquad (5.19)$$

Thus, performing the minimization of Eq. (5.17) is tantamount to subtracting the terms $\gamma_1^-\gamma_4^-/2 = \sum_{i,j,k \in \{0,1\}} |1ij1k\rangle\langle 1ij1k|/2$ and $\gamma_5^-\gamma_4^-/2 = \sum_{i,j,k \in \{0,1\}} |ijk11\rangle\langle ijk11|/2$ from the projector witness and then adding the terms which have been subtracted twice in this way, namely $\gamma_1^-\gamma_4^-\gamma_5^-/2$. This results in the witness given in Eq. (5.11).

State No. 11 (H_6) — Similarly, state No. 11 allows to define four subsets, namely $\mathcal{B}_1 = \{1,4\}$, $\mathcal{B}_2 = \{2,4\}$, $\mathcal{B}_3 = \{1,3\}$ and $\mathcal{B}_4 = \{2,3\}$. Applying Lemma 36 leads to the witness of Eq. (5.12).

5.3 Fully PPT witnesses

In this section, we provide analytical construction methods for fully PPT witnesses of graph states. In Sec. 5.3.1, Lemma 37 gives a method analogous to the fully decomposable witnesses in Lemma 34. An example will be given in Sec. 5.3.1.

As in the last section, we then provide an extended method to construct even better witnesses using the witnesses of Lemma 37. This is done in Sec. 5.3.2, with examples in Sec. 5.3.2. This time, however, the extension is more general and can be applied to a larger family of states. Thus, our main results of this section are Lemmata 37 and 38. Finally, we provide a witness for the 2D cluster state in Sec. 5.3.3 which does not fit into the construction methods presented so far.

As mentioned before, fully PPT witnesses are easier to characterize, since they are fully decomposable witnesses with $P_M = 0$ for all M. This allows for a further generalization of the construction methods presented above — however, only resulting in fully PPT witnesses — and for the construction of a new witness for the 2D cluster state.

5.3.1 Arbitrary graph states

Let us first give the analogon to Lemma 34 for fully PPT witnesses.

Lemma 37. *Given a connected graph state $|G\rangle$. Let $\mathcal{B} = \{\beta_i\}$ be a subset of the set of all qubits such that any two qubits in $\mathcal{B}$ are neither neighbors of each other nor have a neighbor in common. We define $b = |\mathcal{B}|$. Let $\sum_{\vec{s}}$ be the sum over all vectors $\vec{s}$ of length b with elements $s_i = \pm 1$ that contain at least two elements which equal -1, i.e., $\sum_{i=1}^b s_i \leq b - 4$. In this case,*

$$W_G = \frac{1}{2}\mathbb{1} - |G\rangle\langle G| - \sum_{\vec{s}} \left(\frac{1}{2} - \frac{1}{2^{m(\vec{s})}}\right) \prod_{i \in \mathcal{B}} \gamma_i^{s_i} \qquad (5.20)$$

5.3 Fully PPT witnesses

is a fully PPT witness for $|G\rangle$. Here, $m(\vec{s})$ is the number of elements $s_i = -1$ in $\vec{s}$, i.e., $m(\vec{s}) = \left(b - \sum_{i=1}^{b} s_i\right)/2$.

The proof is similar to the proof of Lemma 34, but some parts are easier. We present it in Sec. 5.6.

Examples

2D cluster state — When applying the presented construction to the 2D cluster state of Fig. 5.3, one obtains the witness

$$\begin{aligned}W = \frac{1}{2}\mathbb{1} &- |Cl_{4\times 4}\rangle\langle Cl_{4\times 4}| \\
&- \frac{1}{4}\left(\gamma_1^- \gamma_4^- \gamma_{10}^+ \gamma_{16}^+ + \gamma_1^- \gamma_4^+ \gamma_{10}^+ \gamma_{16}^- + \gamma_1^+ \gamma_4^+ \gamma_{10}^- \gamma_{16}^- \right.\\
&\qquad\left. + \gamma_1^+ \gamma_4^- \gamma_{10}^+ \gamma_{16}^+ + \gamma_1^+ \gamma_4^- \gamma_{10}^+ \gamma_{16}^- + \gamma_1^- \gamma_4^+ \gamma_{10}^- \gamma_{16}^+\right) \\
&- \frac{3}{8}\left(\gamma_1^- \gamma_4^- \gamma_{10}^- \gamma_{16}^+ + \gamma_1^- \gamma_4^- \gamma_{10}^+ \gamma_{16}^- + \gamma_1^- \gamma_4^+ \gamma_{10}^- \gamma_{16}^- + \gamma_1^+ \gamma_4^- \gamma_{10}^- \gamma_{16}^-\right) \\
&- \frac{7}{16}\gamma_1^- \gamma_4^- \gamma_{10}^- \gamma_{16}^- .\end{aligned} \quad (5.21)$$

This witness has a white noise tolerance of $p_\text{tol} = \frac{32768}{51455} \approx 0.637$.

5.3.2 Extended construction method

We can now rewrite Lemma 36 for fully PPT witnesses. Although these witnesses have a smaller white noise tolerance, they can be handled easier analytically, which enabled us to relax the premises of Lemma 36. Therefore, one can apply the new lemma to a larger class of states.

Lemma 38. *Given a connected graph state $|G\rangle$ and m subsets $\mathcal{B}_i$ of its qubits that fulfill the following two conditions:*

(i) No set $\mathcal{B}_i$ contains two qubits that have a neighbor in common.

(ii) No two qubits in $\cup_{i=1}^{m} \mathcal{B}_i$ are neighbors of each other.

Moreover, let W_i be the fully PPT witnesses that one can construct from the subsets $\mathcal{B}_i$ according to Lemma 37. Then,

$$W = \sum_{\vec{k} \in \{0,1\}^n} |\vec{k}\rangle\langle\vec{k}| \min_{i=1,\ldots,m} \langle\vec{k}|W_i|\vec{k}\rangle \quad (5.22)$$

is a fully PPT witness.

The proof of Lemma 38 can be found in Sec. 5.6.

Examples

Linear cluster state — Consider an n-qubit linear cluster state as shown in Fig. 5.2 b). We define a subset $\mathcal{B}_1$ for the construction of a witness W_1 according to Lemma 37) by picking the qubits $\mathcal{B}_1 = \{1, 5, 9, ...\}$. These are marked by circles in Fig. 5.2 b). Analogously, the qubits marked by a square belong to a second subset $\mathcal{B}_2$ which is used to construct a witness W_2. Then, Lemma 38 implies that there is a witness W as given in Eq. (5.22).

Let us present this witness for a seven-qubit cluster state. Then, $\mathcal{B}_1 = \{1, 5\}$ and $\mathcal{B}_2 = \{3, 7\}$. Consequently,

$$W_1 = \frac{1}{2}\mathbb{1} - |Cl_7\rangle\langle Cl_7| - \frac{1}{4}\gamma_1^- \gamma_5^-, \tag{5.23}$$

$$W_2 = \frac{1}{2}\mathbb{1} - |Cl_7\rangle\langle Cl_7| - \frac{1}{4}\gamma_3^- \gamma_7^-. \tag{5.24}$$

Since the only terms that $\gamma_1^- \gamma_5^-$ and $\gamma_3^- \gamma_7^-$ have in common are given by $\gamma_1^- \gamma_3^- \gamma_5^- \gamma_7^-$, Eq. (5.22) can be expressed as

$$W_{\mathrm{Cl}7,2} = \frac{1}{2}\mathbb{1} - |Cl_7\rangle\langle Cl_7| - \frac{1}{4}\gamma_1^- \gamma_5^- - \frac{1}{4}\gamma_3^- \gamma_7^- + \frac{1}{4}\gamma_1^- \gamma_3^- \gamma_5^- \gamma_7^-. \tag{5.25}$$

A fully PPT witness for the seven-qubit linear cluster state constructed according to Lemma 37 with $\mathcal{B} = \{1, 4, 7\}$ has a white noise tolerance of $p_{\mathrm{tol}} = 64/109 \approx 0.588$. The witness of Eq. (5.25), however, only has a tolerance of $p_{\mathrm{tol}} = 64/113 \approx 0.566$. While Lemma 38 does not allow to construct more robust witnesses for linear cluster states compared to simply using Lemma 37, it still has some advantages.

First, for many graph states, e.g. the state No. 6 (Y_5) and the state No. 11 (H_6) of Fig. 2.2, Lemma 38 *does* provide a method to construct witnesses that are more robust than witnesses constructed via Lemma 37 alone. We note that the fully decomposable witnesses of Lemma 36 are even more robust. However, as mentioned before, the prerequisites for Lemma 36 are more strict than those for Lemma 38 and therefore, there are graph states for which the former cannot be used, but the latter applies. For example, this is the case for the 2D cluster state of 16 qubits, to which Lemma 38 can be applied, as we will see at the end of this section, but Lemma 36 can not be used as there are no two qubits with the same neighborhood.

Second, witnesses constructed according to Lemma 38 using two sets $\mathcal{B}_1$ and $\mathcal{B}_2$ as shown in Fig. 5.2 b) can be used to improve the linear cluster state witnesses $\mathcal{W}^{(C_N)}$ of Ref. [72], which results in a witness that only needs two experimental settings to be measured.

To illustrate this, we consider the seven-qubit linear cluster state and its witness $W_{\mathrm{Cl}7,2}$ of Eq. (5.25) again. The linear cluster state witness of Eq. (9) in Ref. [72] is given by

$$\mathcal{W}^{(C_N)} = \frac{3}{2}\mathbb{1} - \left(\prod_{i=1,3,5,7} \gamma_i^+ + \prod_{i=2,4,6} \gamma_i^+\right). \tag{5.26}$$

5.3 Fully PPT witnesses

Due to the form of the generators, it can be measured locally using only two settings, namely the eigenbases of $X_1Z_2X_3Z_4X_5Z_6X_7$ and $Z_1X_2Z_3X_4Z_5X_6Z_7$. Since $\mathcal{W}^{(C_N)} \geq \frac{1}{2}\mathbb{1} - |G\rangle\langle G|$, one has

$$\begin{aligned}
W_{Cl7,2} &= \frac{1}{2}\mathbb{1} - |Cl_7\rangle\langle Cl_7| - \frac{1}{4}\gamma_1^-\gamma_5^- - \frac{1}{4}\gamma_3^-\gamma_7^- + \frac{1}{4}\gamma_1^-\gamma_3^-\gamma_5^-\gamma_7^- \\
&\leq \mathcal{W}^{(C_N)} - \frac{1}{4}\gamma_1^-\gamma_5^- - \frac{1}{4}\gamma_3^-\gamma_7^- + \frac{1}{4}\gamma_1^-\gamma_3^-\gamma_5^-\gamma_7^- \\
&= \mathcal{W}^{(C_N)}_{\text{imp}},
\end{aligned} \quad (5.27)$$

where the last equality sign defines the improved witness $\mathcal{W}^{(C_N)}_{\text{imp}}$. This witness detects more states than $\mathcal{W}^{(C_N)}$ and also requires only two settings, since the additional terms can be determined through the measurement of $X_1Z_2X_3Z_4X_5Z_6X_7$. Note that this is not in contradiction with the result of Ref. [73] stating that $\mathcal{W}^{(C_N)}$ has the highest possible white noise tolerance amongst all stabilizer witnesses that can be measured using two settings, as only witnesses obeying $\mathcal{W}^{(C_N)} \geq \alpha(\frac{1}{2}\mathbb{1} - |G\rangle\langle G|)$ for some $\alpha > 0$ where considered in Ref. [73].

Note that it is possible to construct a better witness for linear cluster state of seven qubits by adding a third witness W_3 constructed for the subset $\mathcal{B}_3 = \{1,7\}$. Then, the white noise tolerance increases to $p_{\text{tol}} = \frac{64}{111} \approx 0.577$.

Finally, we apply the construction of Lemma 38 to the 2D cluster state of 16 qubits.

2D cluster state — Fig. 5.3 b) shows how to choose four subsets $\mathcal{B}_i$ of qubits from a 2D cluster state $|Cl_{4\times 4}\rangle$ made up of 16 qubits. $\mathcal{B}_1$ is shown by circles, $\mathcal{B}_2$ by triangles, $\mathcal{B}_3$ by squares and $\mathcal{B}_4$ by pentagons. The resulting witnesses W_i can be combined as in Eq. (5.22) to yield a witness that can be rewritten as

$$\begin{aligned}
W = &\frac{1}{2}\mathbb{1} - |Cl_{4\times 4}\rangle\langle Cl_{4\times 4}| \\
&-\frac{1}{4}\left(\gamma_1^-\gamma_{11}^- + \gamma_6^-\gamma_{16}^- + \gamma_3^-\gamma_9^- + \gamma_8^-\gamma_{14}^-\right) \\
&+\frac{1}{4}\left(\gamma_1^-\gamma_{11}^-\gamma_6^-\gamma_{16}^- + \gamma_6^-\gamma_{16}^-\gamma_3^-\gamma_9^- + \gamma_3^-\gamma_9^-\gamma_8^-\gamma_{14}^- \right. \\
&\qquad \left. + \gamma_1^-\gamma_{11}^-\gamma_8^-\gamma_{14}^- + \gamma_6^-\gamma_{16}^-\gamma_8^-\gamma_{14}^- + \gamma_1^-\gamma_{11}^-\gamma_3^-\gamma_9^-\right) \\
&-\frac{1}{4}\left(\gamma_1^-\gamma_{11}^-\gamma_6^-\gamma_{16}^-\gamma_3^-\gamma_9^- + \gamma_1^-\gamma_{11}^-\gamma_6^-\gamma_{16}^-\gamma_8^-\gamma_{14}^- \right. \\
&\qquad \left. + \gamma_1^-\gamma_{11}^-\gamma_3^-\gamma_9^-\gamma_8^-\gamma_{14}^- + \gamma_6^-\gamma_{16}^-\gamma_3^-\gamma_9^-\gamma_8^-\gamma_{14}^-\right) \\
&+\frac{1}{4}\gamma_1^-\gamma_{11}^-\gamma_6^-\gamma_{16}^-\gamma_3^-\gamma_9^-\gamma_8^-\gamma_{14}^- .
\end{aligned} \quad (5.28)$$

This witness has a white noise tolerance of $p_{\text{tol}} = \frac{32768}{54335} \approx 0.603$. As we noted for linear cluster state, there are even more subsets $\mathcal{B}_i$ that one can use, such as $\mathcal{B}_5 = \{1,8\}$ and $\mathcal{B}_6 = \{3,9,16\}$. In fact, there are 13 subsets of $\{1,3,6,8,9,11,14,16\}$ that obey condition (ii) of Lemma 38. Taking all of them into account, one obtains a witness with white noise tolerance $p_{\text{tol}} = \frac{32768}{49791} \approx 0.658$ which is even better than the witness of Eq. (5.21).

5.3.3 2D cluster state

Finally, we present a fully PPT witness for the 2D cluster state $|Cl_{4\times 4}\rangle$ of 16 qubits which does not fit into the framework of Lemma 37. Although the construction can easily be generalized to $n \times n$ qubits, we present the witness for the 4×4 case here. To circumvent any problems that might occur due to the border, we consider this state on a torus, i.e., with periodic boundary conditions as shown in Fig. 5.4.

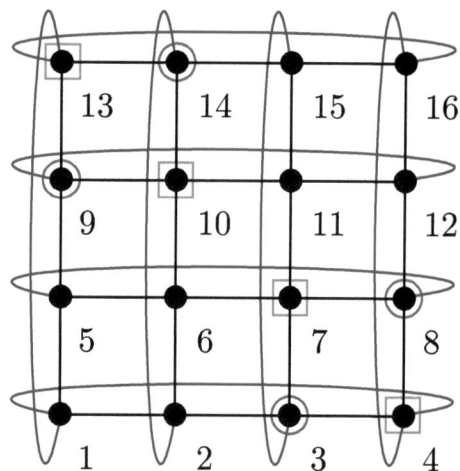

Figure 5.4: For the 2D cluster state on a torus, it is possible to define a fully PPT witness using the diagonals (cf. Lemma 39).

The 2D cluster state has four parallel diagonals in one direction and, orthogonal to these, another set of four diagonals. All of these diagonals contain four qubits. The first set is made up of diagonals parallel to the diagonal $\{3, 8, 9, 14\}$ which is indicated by circles. We denote this set by

$$\mathcal{D}_/ = \{\mathcal{D}_/^{(j)}\} = \{\{1, 6, 11, 16\}, \{2, 7, 12, 13\}, \{3, 8, 9, 14\}, \{4, 5, 10, 15\}\}. \tag{5.29}$$

The second set contains diagonals parallel to the one marked by squares, $\{4, 7, 10, 13\}$. We define it as

$$\mathcal{D}_\backslash = \{\mathcal{D}_\backslash^{(j)}\} = \{\{1, 8, 11, 14\}, \{2, 5, 12, 15\}, \{3, 6, 9, 16\}, \{4, 7, 10, 13\}\}. \tag{5.30}$$

5.4 Entanglement monotone

We can now introduce the following witness.

Lemma 39. *Given the 2D cluster state of 16 qubits with periodic boundary conditions $|Cl_{4\times 4}\rangle$. By $\mathcal{D}_/$ and $\mathcal{D}_\backslash$, we denote the two sets of diagonals as defined above. For each pair of orthogonal diagonals that have no qubit in common, i.e. for each (i,j) such that $\mathcal{D}_/^{(i)} \cap \mathcal{D}_\backslash^{(j)} = \{\}$, we define a projector*

$$D_{(i,j)} = \frac{1}{2}(\mathbb{1} - \prod_{k \in \mathcal{D}_/^{(i)}} g_k) \frac{1}{2}(\mathbb{1} - \prod_{l \in \mathcal{D}_\backslash^{(j)}} g_l). \tag{5.31}$$

Then,

$$W_{4\times 4} = \frac{1}{2}\mathbb{1} - |Cl_{4\times 4}\rangle\langle Cl_{4\times 4}| - \frac{1}{4}\sum_{\vec{k}} |\vec{k}\rangle\langle\vec{k}| \max_{(i,j)} \langle\vec{k}|D_{(i,j)}|\vec{k}\rangle \tag{5.32}$$

is a fully PPT witness for $|Cl_{4\times 4}\rangle$.

The proof can be found in Sec. 5.6. Note that Eq. (5.32) is easy to generalize to $n \times n$ qubits, as for a larger number of qubits only the definitions of Eqs. (5.29) and (5.30) would have to be changed. The proof provided in Sec. 5.6 works for $n \times n$ qubits with $n \geq 3$.

Specifically, the maximization in Eq. (5.32) is carried out over the operators $D_{(1,2)}$, $D_{(1,4)}$, $D_{(2,3)}$, $D_{(2,1)}$, $D_{(3,2)}$, $D_{(3,4)}$, $D_{(4,3)}$ and $D_{(4,1)}$. Similarly to Eq. (5.28), this maximum can also be written as a polynomial in the operators $D_{(i,j)}$. Moreover, the expectation values of these operators can be determined by measuring one experimental setting, namely X-measurements on all qubits. Thus, the sum in Eq. (5.32) can be obtained by implementing one experimental setting.

In order to determine the terms that the witness $W_{4\times 4}$ contains in addition to the projector witness, i.e. the sum in Eq. (5.32), one has to measure the operators $D_{(i,j)}$ that obey $\mathcal{D}_/^{(i)} \cap \mathcal{D}_\backslash^{(j)} = \{\}$. These are the operators $D_{(1,2)}$, $D_{(1,4)}$, $D_{(2,3)}$, $D_{(2,1)}$, $D_{(3,2)}$, $D_{(3,4)}$, $D_{(4,3)}$ and $D_{(4,1)}$. From these, one can determine the elementwise maximum in Eq. (5.32), as one can show that it can be written as a polynomial in the operators $D_{(i,j)}$. Moreover, the expectation values of all of these operators can be determined by measuring one experimental setting, namely X-measurements on all qubits. Thus, the additional term in Eq. (5.32) can be obtained by implementing one experimental setting.

The white noise tolerance of $W_{4\times 4}$ is given by $p_{\text{tol}} = \frac{32768}{53503} \approx 0.612$.

5.4 Entanglement monotone

Finally, we consider the monotone $N(\varrho)$ of Eq. (4.4) and derive the value that it takes for graph states.

Lemma 40. *For any state ϱ of n qubits,*

$$N(\varrho) \leq \frac{1}{2}. \tag{5.33}$$

For any connected graph state $|G\rangle$,

$$N(|G\rangle\langle G|) = \frac{1}{2}. \qquad (5.34)$$

Therefore, connected graph states are maximally entangled states for this monotone. We note that, if the system does not only consist of qubits, but also of higher-dimensional particles, Eq. (5.33) must be replaced by

$$N(\varrho) \leq \frac{1}{2}(d_{\min} - 1), \qquad (5.35)$$

where $d_{\min}$ is the lowest dimension of any particle in the system.

The proofs of this section are given in Sec. 5.6.

5.5 Conclusion

In this chapter we presented general construction methods for graph state witnesses in the framework of PPT mixtures [92]. These methods can be applied to a large class of graph states, resulting in witnesses that are significantly better than previously known witnesses. In many cases, the white noise tolerances approach one for an increasing particle number. This means that for many qubits, the state fidelity can decrease exponentially, but still entanglement is present and can be detected. Moreover, the improvement of the witnesses comes with very low experimental costs, as the additional terms which are not part of the standard projector witness can be measured with one local setting.

For these reasons, we believe that the presented entanglement witnesses will prove to be useful in experiments, also for future experiments involving larger qubit numbers. Furthermore, the applied methods can serve as starting points for the construction of even better entanglement criteria.

There are several interesting questions remaining. First, as we have seen, the approach of this and the last chapter results in strong separability conditions for noisy graph states. A natural question would be whether these conditions are already necessary sufficient for entanglement, or whether they can still be improved. We will tackle this question in the following Sec. 6.

Second, there are many other interesting families of multi-qubit states besides graph states, in particular Dicke states and other states with many symmetries, such as permutation-invariant states. It would be desirable to similarly develop witnesses for these families of states using the framework developed here.

5.6 Proofs

Linear program for graph-diagonal states (Lemma 33)

Proof. Let us define a simplifying notation for this proof: For any operator O we define its graph-diagonal form as $\overline{O} = \sum_{\vec{k}} |\vec{k}\rangle\langle\vec{k}|O|\vec{k}\rangle\langle\vec{k}|$. Note that any state ϱ can be transformed into its graph-diagonal form $\overline{\varrho}$ by local operations. Now suppose that the operator W is the fully decomposable

5.6 Proofs

entanglement witnesses that minimizes the expectation value for the graph diagonal state ϱ_G according to the original problem of Eq. (4.3). Then its graph-diagonal operator $\overline{W}$ has the same expectation value $\text{Tr}(W\rho_G) = \text{Tr}(\overline{W}\rho_G)$ as the original witness. Given any valid decomposition $W = P + Q^{T_M}$ for a particular chosen bipartition M, the operator $\overline{W} = \overline{P} + \overline{Q^{T_M}}$ can be expressed in its corresponding graph-diagonal operators $\overline{P}$ and $\overline{Q^{T_M}}$ due to linearity, but note that $\overline{Q^{T_M}}$ stands for the graph-diagonal form of the partially transposed operator.

However, it is straightforward to see that this operator is actually identical to the partial transpose of the graph-diagonal operator, i.e., $\overline{Q^{T_M}} = \overline{Q}^{T_M}$, as follows: The mapping of $Q \mapsto \overline{Q}$ is achieved by expanding O in the Pauli basis, $Q = \sum_{\vec{x} \in \{0,1,2,3\}^n} \alpha_{\vec{x}} \otimes_{i=1}^{n} \sigma_{x_i}$, and then setting to zero all coefficients $\alpha_{\vec{x}}$ of Pauli matrix products which are no stabilizers of the given graph state. Note that σ_1, σ_2, σ_3 denote the Pauli matrices and σ_0 is the identity. In this picture, the partial transposition only corresponds to flipping the sign of coefficients $\alpha_{\vec{x}}$ of Pauli matrix products which change under partial transposition. These are the Pauli matrix products in which there is an odd number of σ_2s, i.e. of Ys, in the set M, since $Y^T = -Y$ and all other Pauli matrices are invariant under transposition.

Then, it is clear that the partial transposition and the mapping $Q \mapsto \overline{Q}$ commute. Thus the witness decomposition simplifies to $\overline{W} = \overline{P} + \overline{Q}^{T_M}$. Since the operator $P \geq 0$ is positive semidefinite, the overlap with any basis element $\langle \vec{k}|P|\vec{k}\rangle \geq 0$ is non-negative. However this is equivalent to $\overline{P} \geq 0$ because $\overline{P}$ is diagonal in exactly this basis. The same argument applies to the operator Q, which concludes the proof. □

Fully decomposable witnesses for arbitrary graph states (Lemma 34)

Proof. Consider an arbitrary, connected graph $G = (V, E)$ consisting of a set V of vertices/qubits and a set E of edges that connect some of these vertices.

In the following, $\mathcal{N}(i) = \mathcal{N}(i) \cup \{i\}$ denotes the union of qubit i and its neighborhood. Moreover, all states in the following are given in the graph state basis of the corresponding graph.

Let us first give four lemmata to prepare the main proof. The first of these lemmata shows which kind of partial transposition one can apply to one of two orthogonal vectors without affecting their orthogonality. The second one can be used to estimate the eigenvalues of a partially transposed state. More precisely, it provides an upper bound on these eigenvalues in terms of the state's Schmidt coefficients. The third lemma demonstrates that certain expressions are invariant under partial transpositions on a single qubit. Finally, the fourth lemma helps to estimate the largest Schmidt coefficient of a graph state. In order to prove it, we will count the Bell pairs that can be distilled from it using local operations and classical communication (LOCC).

We will then apply these lemmata to prove that the operator W_G of Eq. (5.14) is a fully decomposable witness.

Lemma 41. *Given a graph $G = (V, E)$ of n qubits and an arbitrary bipartition $M|\overline{M}$ of these qubits. Let $|\vec{a}\rangle$ and $|\vec{c}\rangle$ be two arbitrary states in the associated graph state basis. If there is a qubit i with*

77

$\tilde{\mathcal{N}}(i) \subseteq M$ or $\tilde{\mathcal{N}}(i) \subseteq \overline{M}$, such that $c_i \neq a_i$, then

$$\langle \vec{c}| \left(|\vec{a}\rangle\langle\vec{a}|\right)^{T_M} |\vec{c}\rangle = 0 \,. \tag{5.36}$$

Proof. Let $g_i, i = 1\ldots n$ be the generators defined by Eq. (2.57). Since $X^T = X$, $Y^T = -Y$, $Z^T = Z$ and $\mathbb{1}^T = \mathbb{1}$, the partial transposition of a product of generators only changes the product's sign. Thus, we can describe the action of the partial transpose T_M w.r.t partition M on products of generators by

$$\left(\prod_{i=1}^n g_i^{x_i}\right)^{T_M} = (-1)^{f(\vec{x})} \left(\prod_{i=1}^n g_i^{x_i}\right), \tag{5.37}$$

where $x_i \in \{0,1\}$. Here, f depends on M and is a Boolean function defined by

$$f: \{0,1\}^n \to \{0,1\} \tag{5.38}$$

$$\vec{x} \mapsto f(\vec{x}) = \begin{cases} 0, & \text{if } \left(\prod_{i=1}^n g_i^{x_i}\right)^{T_M} = \prod_{i=1}^n g_i^{x_i} \\ 1, & \text{if } \left(\prod_{i=1}^n g_i^{x_i}\right)^{T_M} = -\prod_{i=1}^n g_i^{x_i} \end{cases}.$$

Note that the *support* $\text{supp}(f)$ of a Boolean function contains the bits that the function depends on, i.e.,

$$\text{supp}(f) = \{i \mid \exists\, \vec{x}, \text{s.t.} f(x_1,\ldots,x_i,\ldots,x_n) \neq f(x_1,\ldots,x_i \oplus 1,\ldots,x_n)\}\,. \tag{5.39}$$

Due to the explicit form of the g_i, flipping the value of x_i cannot change $f(\vec{x})$, if $i \in \mathcal{U}_M$. Therefore, $\mathcal{U}_M \cap \text{supp}(f) = \{\}$. For this reason, we can pull qubits in $\mathcal{U}_M$ out of the partial transposition T_M in the following way (using also Eq. (2.69)).

$$\langle \vec{b}| \left(|\vec{a}\rangle\langle\vec{a}|\right)^{T_M} |\vec{b}\rangle = \text{Tr}\left[\prod_{j=1}^n \frac{(-1)^{b_j} g_j + \mathbb{1}}{2} \prod_{i \in \mathcal{U}_M} \frac{(-1)^{a_i} g_i + \mathbb{1}}{2} \left(\prod_{i \notin \mathcal{U}_M} \frac{(-1)^{a_i} g_i + \mathbb{1}}{2}\right)^{T_M}\right] \tag{5.40}$$

Since $\frac{g_i + \mathbb{1}}{2} \cdot \frac{-g_i + \mathbb{1}}{2} = 0$, the last expression vanishes if there is an $i \in \mathcal{U}_M$, such that $b_i \neq a_i$. $\square$

The states in the following lemma are generic states and no graph state basis vectors.

Lemma 42. *Given a state $|\psi\rangle$ and its Schmidt decomposition $|\psi\rangle = \sum_{i=1}^{d_1} \lambda_i |\mu_i\rangle \otimes |\nu_i\rangle$ with respect to some bipartition $M|\overline{M}$, where $\lambda_i \geq 0$, $d_1 = \dim(M)$, $d_2 = \dim(\overline{M})$ and w.l.o.g. $d_1 \leq d_2$. Then, for any state $|\phi\rangle$,*

$$\langle\phi| \left(|\psi\rangle\langle\psi|\right)^{T_M} |\phi\rangle \leq \max_i \lambda_i^2 \,. \tag{5.41}$$

Proof. Writing down $(|\psi\rangle\langle\psi|)^{T_M}$ in the basis $\{|\mu_i\rangle \otimes |\nu_j\rangle\}_{i=1\ldots d_1, j=1\ldots d_2}$, one obtains a matrix with two different kinds of submatrices. First, a diagonal one with diagonal elements λ_i^2 or zero. Second,

5.6 Proofs

anti-diagonal submatrices of the form

$$\begin{pmatrix} 0 & \lambda_i \lambda_j \\ \lambda_i \lambda_j & 0 \end{pmatrix}. \tag{5.42}$$

Thus, the eigenvalues of the total matrix are $\{\pm \lambda_i \lambda_j, \lambda_i^2, 0\}$ and the maximum of these eigenvalues has the form λ_i^2. □

Let us now return to the graph state basis and recall that the application of the Pauli operator Z_k to a graph state basis vector results in a bit flip on bit k, i.e.,

$$Z_k|\vec{a}\rangle = |a_1 \ldots a_{k-1}\, a_k \oplus 1\, a_{k+1} \ldots a_n\rangle. \tag{5.43}$$

Lemma 43. *Given a graph G. Then, in the associated graph state basis,*

$$(|\vec{a}\rangle\langle\vec{a}| + |\vec{c}\rangle\langle\vec{c}|)^{T_k} = |\vec{a}\rangle\langle\vec{a}| + |\vec{c}\rangle\langle\vec{c}|, \tag{5.44}$$

i.e. $|\vec{a}\rangle\langle\vec{a}| + |\vec{c}\rangle\langle\vec{c}|$ is invariant under partial transposition on qubit k, if

$$|\vec{c}\rangle = \prod_{i \in \mathcal{N}(k)} Z_i |\vec{a}\rangle. \tag{5.45}$$

Proof. With Eq. (2.69) and Eq. (5.43), we have

$$|\vec{a}\rangle\langle\vec{a}| + |\vec{b}\rangle\langle\vec{b}|$$

$$= \left(\frac{g_k + \mathbb{1}}{2} + \frac{-g_k + \mathbb{1}}{2}\right) \prod_{\substack{i=1 \\ i \neq k}}^{n} \frac{(-1)^{a_i} g_i + \mathbb{1}}{2}$$

$$= \prod_{\substack{i=1 \\ i \neq k}}^{n} \frac{(-1)^{a_i} g_i + \mathbb{1}}{2}. \tag{5.46}$$

Since g_k cancels in Eq. (5.46), the explicit form of the generators g_i implies that, in this equation, there is no Y on qubit k. Since Y is the only Pauli matrix that changes under partial transposition, $|\vec{a}\rangle\langle\vec{a}| + Z_k|\vec{a}\rangle\langle\vec{a}|Z_k$ is invariant under T_k. □

Lemma 44. *Let $|G\rangle$ be a graph state that is defined by a bipartite graph $G = (V, E)$, i.e. the qubits can be grouped into two partitions M and $\overline{M}$, such that no two qubits in the same partition are connected with each other. Let λ_i be the Schmidt coefficients of $|G\rangle$ with respect to the bipartition $M|\overline{M}$. If there exists a subset $\mathcal{B} = \{\beta_i\}$ of m qubits which have at least one neighbor and are chosen*

5 Entanglement witnesses for graph states

in such a way that no two qubits in $\mathcal{B}$ have a neighbor in common or are neighbors of each other, then

$$\max_i \lambda_i^2 \leq 2^{-m} . \tag{5.47}$$

Proof. Note that any graph can be made bipartite with respect to a fixed bipartition $M|\overline{M}$ using operations which are local with respect to $M|\overline{M}$. These operations are controlled-Z between two qubits i, j of the same partition and they correspond to a deletion of the edge between qubits i and j [77].

In order to prove that the square of the largest Schmidt coefficient of $|G\rangle$ is smaller than (or equal to) 2^{-m}, it is sufficient to show that $|G\rangle$ can be converted into at least m Bell pairs via local operations and classical communication. Since the largest Schmidt coefficient does not decrease under LOCC [110] and a Bell pair has Schmidt coefficients $\{1/\sqrt{2}, 1/\sqrt{2}\}$, this implies the given bound.

In the first step, we choose a set of edges $F = \{(\beta_i, w_i)\} \subseteq E$ by selecting, for every qubit β_i in $\mathcal{B}$, a neighboring w_i. The edge (β_i, w_i) between them then belongs to F. Since no qubit w_i can be a neighbor of two different qubits in $\mathcal{B}$ according to the assumptions, every qubit in the graph is endpoint of at most one of the edges in F. A set with this property is also called a *matching*. For our proof, each edge in the matching F marks two qubits between which we will create a Bell pair which is disconnected from the rest of the graph.

As a second step, we measure every qubit, which is not an end point of an edge in F, in the Z-basis. In terms of the graph, this deletes all edges that are incident on a measured qubits. Fig. 5.5 shows an example of a graph that emerges from these measurements. There are two kinds of edges left: edges that are contained in the matching (shown as thick vertical lines in Fig. 5.5) and edges that connect a qubit w_i to a qubit w_j in the opposite partition, which are not in the matching (drawn thinner and in black). Note that, after the measurements, the qubits β_i are only connected through edges of the matching. Any other edge would either contradict the fact that the graph is bipartite with respect to $M|\overline{M}$ or the condition that qubits in $\mathcal{B}$ have no neighbor in common. As seen in Fig. 5.5, some qubits β_i are in M, some are in $\overline{M}$. This distinction, however, is of no importance in this proof. Also, there might be other, isolated qubits. These are not shown in Fig. 5.5, since they do not play any role in the proof.

Finally, we need to delete all edges that are not in the matching, i.e. the edges (w_i, w_j). Consider an edge, say (w_1, w_k) (cf. Fig. 5.5). It can be deleted using the following steps:

First, connect β_1 and w_k. Such a creation of an edge corresponds to an application of a local unitary to the graph state, namely a controlled-Z gate acting on the two qubits to be connected.

Second, apply a local complementation operation on qubit β_1. This operation corresponds to a local unitary and inverts the neighborhood graph of β_1. More precisely, all edges between neighbors of β_1 are deleted and all neighbors of β_1 which are not connected become connected [77]. Since w_1 and w_k are the only neighbors of β_1, this means that the edge (w_1, w_k) is deleted.

Finally, delete the edge (β_1, w_k) again. The described steps now have to be repeated for all other edges that do not belong to the matching. After that, one ends up with m pairs of connected qubits

5.6 Proofs

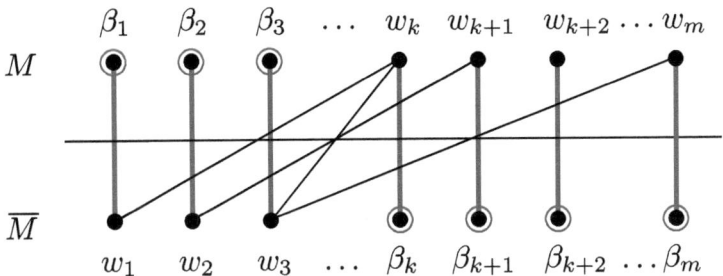

Figure 5.5: After measuring out all qubits that are not needed for the creation of Bell pairs, one obtains a graph as the one shown. Edges of the matching are indicated by thick vertical lines, while other edges are shown with thin lines.

which are disconnected from the rest of the graph. These m Bell pairs have a largest Schmidt coefficient of $\sqrt{2}^{-m}$ and the performed LU and LOCC operations cannot have decreased it [110]. Thus, the square of the largest Schmidt coefficient of $|G\rangle$ must be smaller than 2^{-m}. □

Let us now start with the main part of the proof in which Lemmata 41 - 44 will be used.

For the sake of brevity, we define $P_+ = \sum_{\vec{s}} \prod_{i \in \mathcal{B}} \gamma_i^{s_i}$ [cf. Eq. (5.14)]. Note that P_+ is a sum of all projectors onto graph state basis vectors that contain at least two excitations in $\mathcal{B}$, i.e., two bits β_i that equal one. For example, in the case of a linear cluster state (cf. Fig. 5.2), we can choose $\mathcal{B} = \{1, 4, 7, \dots\}$. Then,

$$P_+ = \sum_{\vec{x} \in \{0,1\}^{n-b}} (|0x_1 x_2 1 x_3 x_4 1 \dots\rangle\langle 0 x_1 x_2 1 x_3 x_4 1 \dots| \\
+ |1 x_1 x_2 0 x_3 x_4 1 \dots\rangle\langle 1 x_1 x_2 0 x_3 x_4 1 \dots| \\
+ |1 x_1 x_2 1 x_3 x_4 0 \dots\rangle\langle 1 x_1 x_2 1 x_3 x_4 0 \dots| \\
+ |1 x_1 x_2 1 x_3 x_4 1 \dots\rangle\langle 1 x_1 x_2 1 x_3 x_4 1 \dots| \\
+ \dots). \tag{5.48}$$

Note that the following proof is an extension of the proof for linear cluster states in Ref. [92].

Main part of the proof of Lemma 34 — In order to prove that W_G is a fully decomposable witness, we have to show that, for every strict subset M, there exists a positive operator P_M such that

$$Q_M = (W_G - P_M)^{T_M} \geq 0. \tag{5.49}$$

We proceed in two steps. First, for a given M, we transform our problem for the graph state $|G\rangle$ into a problem for another graph state $|G'\rangle$ in which some edges have been deleted by local operations.

Second, in the main part of the proof, we provide an algorithm for a given M to construct a positive operator P_M that obeys Eq. (5.49).

First step: Transformation of the graph state — The goal of the first step is to transform the graph G to a graph G' by deleting all edges that connect qubits in the same partition. A graph, in which the vertices can be divided into two subsets M and $\overline{M}$ such that only vertices of different subsets are connected with each other, is called *bipartite*. As we will see later, this property will be useful, since it allows us to make use of Lemma 43.

We start by noting that any operator O that is diagonal in a graph basis can be written in the form

$$O = \sum_{\vec{x}} c_{\vec{x}} \prod_{i=1}^{n} g_i^{x_i}, \tag{5.50}$$

where the sum runs over the set of binary vectors $\vec{x} \in \{0,1\}^n$. Moreover, $c_{\vec{x}}$ are coefficients that depend on the operator O. Since any partial transposition can at most introduce minus signs in some terms of this sum, such operators remain diagonal under any partial transposition.

As both W_G of Eq. (5.14) is graph-diagonal and we restrict ourselves to operators P_M which are also graph-diagonal, it is enough to prove that

$$\langle \vec{k} | (W_G - P_M)^{T_M} | \vec{k} \rangle \geq 0 \tag{5.51}$$

holds for all M and all graph state basis vectors $|\vec{k}\rangle$.

Now, we perform the graph transformation $G \mapsto G'$ by deleting all edges that connect qubits in the same partition. This corresponds to applying a controlled-Z operation $C_{j,l}$ to all such pairs of qubits j,l. Altogether, this results in a unitary $A = \prod_{(j,l)} C_{j,l}$ that acts on M, where the product runs over all edges (j,l) that connect qubits in M, and an analogous unitary $B = \prod_{(j,l)} C_{j,l}$, where the product includes edges in $\overline{M}$ and which acts on $\overline{M}$.

Since the controlled-Z operation is real and diagonal, we have

$$A = A^* = A^\dagger = A^T \tag{5.52}$$

and analogously for B.

Together with the unitarity of A and B, these equalities imply the equivalence

$$\langle \vec{k} | (W_G - P_M)^{T_M} | \vec{k} \rangle \geq 0 \tag{5.53}$$

$$\Leftrightarrow {}_{G'}\langle \vec{k} | A \otimes B \, (W_G - P_M)^{T_M} \, A^\dagger \otimes B^\dagger | \vec{k} \rangle_{G'} \geq 0 \tag{5.54}$$

$$\Leftrightarrow {}_{G'}\langle \vec{k} | A \otimes B^* \, (W_G - P_M)^{T_M} \, A^\dagger \otimes B^T | \vec{k} \rangle_{G'} \geq 0 \tag{5.55}$$

$$\Leftrightarrow {}_{G'}\langle \vec{k} | \left[A \otimes B \, (W_G - P_M) \, A^\dagger \otimes B^\dagger \right]^{T_M} | \vec{k} \rangle_{G'} \geq 0 \tag{5.56}$$

$$\Leftrightarrow {}_{G'}\langle \vec{k} | (W_{G'} - P'_M)^{T_M} | \vec{k} \rangle_{G'} \geq 0 \tag{5.57}$$

5.6 Proofs

where $|\vec{k}\rangle_{G'} = A \otimes B |\vec{k}\rangle$ are the basis vectors that are associated to the transformed generators $g_i' = (A \otimes B)g_i(A^\dagger \otimes B^\dagger)$. Also, the transformed witness is given by $W_{G'} = (A \otimes B)W_G(A^\dagger \otimes B^\dagger) = \frac{1}{2}\mathbb{1} - |G'\rangle\langle G'| - \frac{1}{2}\sum_{\vec{k}^2 > 1} \prod_{i=1}^{|\mathcal{B}|} \frac{1+(-1)^{k_i}g'_{\beta_i}}{2}$.

Thus, the transformed Eq. (5.57) has the same form as Eq. (5.51). Keep in mind that one needs to prove Eq. (5.57) for all subsets M and all basis vectors $|\vec{k}\rangle_{G'}$.

For better readability, we drop the subscript G' of the graph basis vectors: $|\vec{k}\rangle_{G'} \mapsto |\vec{k}\rangle$. Every state in the remainder of this proof is to be understood in the graph basis of graph G'.

Finally, we note that the most important thing to keep in mind from this step is that the graph G' is bipartite with respect to the two sets M and $\overline{M}$.

Second step: Algorithm to construct P_M' — Let us now provide an algorithm to construct P_M' for any given M. Note that we order the qubits β_i in a canonical way such that $\beta_i < \beta_{i+1}$.

1. Start with $P_M^{(0)} = |G'\rangle\langle G'| = |0\ldots 0\rangle\langle 0\ldots 0|$.

2. Set $i = 1$.

3. If β_i has no neighbors (in graph G'), set $P_M^{(i)} = P_M^{(i-1)}$. If β_i has neighbors, define $P_M^{(i)}$ as
$$P_M^{(i)} = P_M^{(i-1)} + \left(\prod_{j \in \mathcal{N}(\beta_i)} Z_j\right) P_M^{(i-1)} \left(\prod_{j \in \mathcal{N}(\beta_i)} Z_j\right).$$

4. If $i \leq b$, increase i by one and repeat step 3. Otherwise, proceed with step 5.

5. Let r be the number of qubits in $\mathcal{B}$ that have neighbors (in graph G'), i.e., the number of steps in which $P_M^{(i)}$ changed.
If $r \leq 1$, define
$$P_M' = 0. \tag{5.58}$$
Let t be the value of i for which $P_M^{(i)}$ was changed the last time, i.e., $P_M^{(i)} = P_M^{(t)} \; \forall \, i > t$. If $r > 1$, define
$$P_M' = P_M^{(t-1)} - |G'\rangle\langle G'|. \tag{5.59}$$

Note that the operator P_M' constructed via the given algorithm is either zero or a sum of one-dimensional projectors onto basis states, i.e.,
$$P_M' = \sum_{\vec{a}} |\vec{a}\rangle\langle\vec{a}|. \tag{5.60}$$

This can be seen by the fact that $P_M^{(0)} = |G'\rangle\langle G'| = |0\ldots 0\rangle\langle 0\ldots 0|$, the application of Z only flips a bit and finally $|G'\rangle\langle G'|$ is subtracted again. Let us illustrate the algorithm by a concrete example.

Example of the algorithm: Consider state No. 16 of Fig. 2.2 and the bipartition given by $M = \{1, 2, 5, 6\}$. Then, the transformation in the first step of the proof deletes the edges $(1, 2)$ and $(3, 4)$, since $1, 2 \in M$ and $3, 4 \in \overline{M}$.

Let us choose set $\mathcal{B} = \{1, 5, 6\}$. Thus, the algorithm produces the following operators. From step 1, we have

$$P_M^{(0)} = |000000\rangle\langle 000000|. \tag{5.61}$$

As qubit 1 does not have any neighbors, since edge $(1,2)$ has been deleted, step 2 does not change the operator $P_M^{(0)}$ and therefore results in

$$P_M^{(1)} = |000000\rangle\langle 000000|. \tag{5.62}$$

Then, the loop in step 3 produces

$$P_M^{(2)} = |000000\rangle\langle 000000| + |000100\rangle\langle 000100|, \tag{5.63}$$

$$P_M^{(3)} = |000000\rangle\langle 000000| + |000100\rangle\langle 000100|$$
$$+ |001000\rangle\langle 001000| + |001100\rangle\langle 001100|. \tag{5.64}$$

$P_M^{(i)}$ was changed in two steps or, in other words, two qubits in graph G' which are also in $\mathcal{B}$, namely qubits 5 and 6, have a neighbor. Thus, $r = 2$. Moreover, as $P_M^{(i)}$ was changed in the third step, we have $t = 3$ and therefore

$$P_M' = P_M^{(2)} - |000000\rangle\langle 000000|$$
$$= |000100\rangle\langle 000100|. \tag{5.65}$$

Therefore, in this example, the sum in Eq. (5.60) has only one term.

Let us now return to the general case and understand the properties of the operator P_M' for an arbitrary M. The construction uses Lemma 43 to ensure that, in every step, either

$$\left(P_M^{(i)}\right)^{T_{M_i}} = \left(P_M^{(i)}\right)^{T_{\widetilde{\mathcal{N}}(\beta_i)}} \tag{5.66a}$$

or

$$\left(P_M^{(i)}\right)^{T_{M_i}} = P_M^{(i)} \tag{5.66b}$$

hold, where we defined $M_k = M \cap \widetilde{\mathcal{N}}(\beta_k)$. Therefore, as we will see later, the qubits β_i can be treated as if they had no neighbor in the opposite partition.

To see that Eqs. (5.66) hold, assume that $\beta_i \in M$. Since qubits that were neighbors of β_i in graph G and were also in M are not connected to β_i in graph G' anymore, we know that $\mathcal{N}(\beta_i) \subseteq \overline{M}$. Then, the given algorithm sets

$$P_M^{(i)} = \sum_{\vec{c}} \left[|\vec{c}\rangle\langle\vec{c}| + \left(\prod_{j \in \mathcal{N}(\beta_i)} Z_j\right)|\vec{c}\rangle\langle\vec{c}| \left(\prod_{j \in \mathcal{N}(\beta_i)} Z_j\right)\right]. \tag{5.67}$$

5.6 Proofs

This expression is invariant under the partial transposition T_{β_i} due to Lemma 43. Therefore, Eq. (5.66b) holds. Similarly, in the case $\beta_i \in \overline{M}$, Eq. (5.66a) holds.

Eqs. (5.66) hold in every step, i.e., for $i = j$ and for $i = k$, where $j \neq k$. According to the premise of non-overlapping neighborhoods of the qubits in $\mathcal{B}$, we have $\widetilde{\mathcal{N}}(\beta_j) \cap \widetilde{\mathcal{N}}(\beta_k) = \{\}$. Therefore, the partial transpositions in Eqs. (5.66) for $i = j$ always affect qubits different from the ones that are affected by the partial transpositions for $i = k$. For this reason, Eqs. (5.66) for $P_M^{(t-1)}$ hold with respect to every value of k, except for $k = t$. More precisely,

$$\left(P_M^{(t-1)}\right)^{T_{M_k}} = \left(P_M^{(t-1)}\right)^{T_{\widetilde{\mathcal{N}}(\beta_k)}} \tag{5.68a}$$

or

$$\left(P_M^{(t-1)}\right)^{T_{M_k}} = P_M^{(t-1)} \tag{5.68b}$$

is true for every $k \neq t$. We will use this important property later.

Let us proceed with the proof. Since P_M' is zero or has the form of Eq. (5.60), we know that $P_M' \geq 0$. Thus, it remains to show that Eq. (5.57) holds.

Note that the transformed operator $P_+' = (A \otimes B) P_+ (A \otimes B)$ is invariant under any partial transposition. This can be seen using Eq. (5.52) and the fact that P_+ is invariant under any partial transposition. $P_+ = \sum_{\vec{s}} \prod_{i \in \mathcal{B}} \gamma_i^{s_i}$ is invariant, since it only contains generators of qubits that have no neighbor in common and are no neighbors of each other. Thus, the form of the generators as given in Eq. (2.57) implies that P_+ does not contain any Y operators which are the only Pauli matrices that change under transposition.

Together with the explicit form of the witness given in Eq. (5.14), we can therefore rewrite Eq. (5.57) as

$$\frac{1}{2} - \frac{1}{2}\langle \vec{k}|P_+'|\vec{k}\rangle - \langle \vec{k}| \left(|G'\rangle\langle G'| + P_M'\right)^{T_M} |\vec{k}\rangle \geq 0 \, . \tag{5.69}$$

In order to prove this, we distinguish two different cases:

1. $\boxed{\langle \vec{k}|P_+'|\vec{k}\rangle \neq 0 \Leftrightarrow \langle \vec{k}|P_+'|\vec{k}\rangle = 1}$

   Note that this equivalence is due to the form of P_+' as shown in Eq. (5.48). Also, this form implies that, in the vectors $|\vec{k}\rangle$ with non-zero overlap, there must be at least two qubits $i_0, j_0 \in \mathcal{B}$, with $i_0 \neq j_0$, such that $k_{i_0} = k_{j_0} = 1$.

   In the case $P_M' = 0$, Eq. (5.69) and $\langle \vec{k}|P_+'|\vec{k}\rangle = 1$ are equivalent to

$$-\langle \vec{k}| \left(|G'\rangle\langle G'|\right)^{T_M} |\vec{k}\rangle \geq 0 \, . \tag{5.70}$$

   To see that the left-hand side always vanishes for all M and all $|\vec{k}\rangle$, one uses that $P_M' = 0$ is equivalent to $r \leq 1$, i.e., $\widetilde{\mathcal{N}}(\beta_i) \subseteq M$ or $\widetilde{\mathcal{N}}(\beta_i) \subseteq \overline{M}$ holds for all qubits $\beta_i \in \mathcal{B}$ with at most

85

one exception, namely β_t. With $k_{i_0} = k_{j_0} = 1$, Lemma 1 can be applied to see that the left-hand side of Eq. (5.70) vanishes.

In the case $P'_M \neq 0$, Eq. (5.69) can be simplified using $\langle \vec{k} | P'_+ | \vec{k} \rangle = 1$ to

$$-\langle \vec{k} | \left(|G'\rangle\langle G'| + P'_M \right)^{T_M} |\vec{k}\rangle \geq 0$$
$$\Leftrightarrow -\langle \vec{k} | \left(P_M^{(t-1)} \right)^{T_M} |\vec{k}\rangle \geq 0 \,. \tag{5.71}$$

Here, the definition of P'_M, Eq. (5.59), has been used.

Now, P'_M and therefore $P_M^{(t-1)}$ consists of a sum of projectors onto graph basis states $|\vec{a}\rangle$ [see Eq. (5.60)]. Since the algorithm starts with $P_M^{(0)} = |0\ldots 0\rangle\langle 0\ldots 0|$ and never flips any bits on the qubits $\beta_i \in \mathcal{B}$, these states $|\vec{a}\rangle$ obey $a_{\beta_i} = 0$, $\forall\, i = 1, \ldots, b$. Also, depending on whether $i_0 = t$ or $j_0 = t$, Eqs. (5.68) can be applied to whichever of these two qubits is different from t. Let us assume that $i_0 \neq t$. Then, one can use Eq. (5.68a) or Eq. (5.68b) to replace M by a slightly modified subset M' with $\widetilde{\mathcal{N}}(\beta_{i_0}) \subseteq M'$ or $\widetilde{\mathcal{N}}(\beta_{i_0}) \subseteq \overline{M'}$, respectively. Thus, Lemma 41 applied to i_0 yields

$$\langle \vec{k} | \left(P_M^{(t-1)} \right)^{T_M} |\vec{k}\rangle = \langle \vec{k} | \left(P_M^{(t-1)} \right)^{T_{M'}} |\vec{k}\rangle$$
$$= 0 \tag{5.72}$$

and therefore Eq. (5.71) holds.

2. $\boxed{\langle \vec{k} | P'_+ | \vec{k} \rangle = 0}$

To show that Eq. (5.69) holds, we need to prove that

$$\langle \vec{k} | \left(|G'\rangle\langle G'| + P'_M \right)^{T_M} |\vec{k}\rangle \leq \frac{1}{2} \,. \tag{5.73}$$

In the case $P'_M \neq 0$, P'_M is given by Eq. (5.60) and Eq. (5.73) is equivalent to

$$\langle \vec{k} | \left(|G'\rangle\langle G'| + \sum_{\vec{a}} |\vec{a}\rangle\langle \vec{a}| \right)^{T_M} |\vec{k}\rangle \leq \frac{1}{2} \,. \tag{5.74}$$

Note that $|G'\rangle\langle G'| + \sum_{\vec{a}} |\vec{a}\rangle\langle \vec{a}| = P_M^{(t-1)}$ consists of 2^{r-1} terms, as one starts with one term and doubles this number $(r-1)$ times to obtain $P_M^{(t-1)}$. Therefore, it is enough to prove the upper

5.6 Proofs

bounds

$$\langle \vec{k}| (|G'\rangle\langle G'|)^{T_M} |\vec{k}\rangle \leq 2^{-r} \tag{5.75}$$

and

$$\langle \vec{k}| (|\vec{a}\rangle\langle \vec{a}|)^{T_M} |\vec{k}\rangle \leq 2^{-r} \; \forall \; |\vec{a}\rangle. \tag{5.76}$$

We will show these bounds using Lemma 42. However, since the vectors $|\vec{a}\rangle$ are basis vectors of the graph state basis of $|G'\rangle$, $|\vec{a}\rangle$ and $|G'\rangle$ are LU-equivalent. Therefore, they have the same Schmidt coefficients and Lemma 42 results in the same upper bounds. For this reason, it suffices to show only one of these upper bounds, namely

$$\langle \vec{k}| (|G'\rangle\langle G'|)^{T_M} |\vec{k}\rangle \leq 2^{-r}. \tag{5.77}$$

In order to apply Lemma 42, we need the largest Schmidt coefficient of $|G'\rangle$. According to Lemma 44, the largest Schmidt coefficient is smaller than (or equal to) 2^{-r}, since r is the number of qubits in $\mathcal{B}$ that have at least one neighbor. Note that the conditions of Lemma 44 are met, since G' is a bipartite graph. Thus, Eq. (5.77) holds.

In the case $P'_M = 0$, we need to show that

$$\langle \vec{k}| (|G'\rangle\langle G'|)^{T_M} |\vec{k}\rangle \leq \frac{1}{2}. \tag{5.78}$$

Since $M \neq \{1, \ldots, n\}$, there is at least one Bell pair in G that connects a qubit in M and a qubit in $\overline{M}$. Since the transformation $G \to G'$ only deletes connections between qubits in the same partition, this pair is also connected in graph G'. Then, however, deleting all edges besides the one of this pair by measuring all other qubits leads to one Bell pair. One Bell pair with Schmidt coefficients $\{\frac{1}{\sqrt{2}}, \frac{1}{\sqrt{2}}\}$ is enough to show that Eq. (5.78) holds (using Lemma 42).

This finishes the proof of Lemma 34.

□

White noise tolerance of fully decomposable witnesses (Corollary 35)

Proof. The definition of the white noise tolerance p_{tol} for state $|G\rangle$ and witness W implies that

$$p_{\text{tol}} = \left[1 - \frac{\text{Tr}(W)}{2^n \langle G|W|G\rangle}\right]^{-1}. \tag{5.79}$$

Since $\langle G|W|G\rangle = -1/2$, it remains to calculate

$$\text{Tr}(W) = 2^{n-1} - 1 - \frac{1}{2} 2^{n-|\mathcal{B}|} \sum_{j=2}^{|\mathcal{B}|} \binom{|\mathcal{B}|}{j}$$
$$= 2^{n-1} - 1 - 2^{n-|\mathcal{B}|-1} \left(2^{|\mathcal{B}|} - |\mathcal{B}| - 1 \right) \tag{5.80}$$

Together with Eq. (5.79), this results in Eq. (5.15). □

Extended construction of fully decomposable witnesses (Lemma 36)

Before we begin with the proof of Lemma 36, let us state the following lemma, which we will need later in this proof and also in Sec. 5.6.

Lemma 45. *Given a graph state $|G\rangle$ of n qubits, the associated generators g_i and the projectors $\gamma_i^{\pm} = (\mathbb{1} \pm g_i)/2$. Let $\mathcal{B}$ be any subset of all qubits in which no two qubits are neighbors of each other. Let $\mathcal{B}_i$ for $i = 1, \ldots, m$, be some arbitrary subsets of $\mathcal{B}$ and P_i for $i = 1, \ldots, m$, some operators that can be written as*

$$P_i = \sum_{\vec{s}} \alpha_{\vec{s}} \prod_{j \in \mathcal{B}_i} \gamma_j^{s_j}, \tag{5.81}$$

where $\sum_{\vec{s}}$ sums over some subset of $\{-1, +1\}^{|\mathcal{B}_i|}$, i.e. over vectors of length $|\mathcal{B}_i|$ with elements ± 1, and $\alpha_{\vec{s}}$ are some coefficients. Then, the operator

$$\max_{i=1,\ldots,m} P_i = \sum_{\vec{k}} |\vec{k}\rangle\langle\vec{k}| \max_{i=1,\ldots,m} \langle\vec{k}|P_i|\vec{k}\rangle \tag{5.82}$$

is invariant under any partial transposition.

Proof. We prove the invariance by showing that $\max_{i=1,\ldots,m} P_i$ can be written as a linear combination of operators

$$T_{\vec{s}} = \prod_{j \in \mathcal{B}} \gamma_j^{s_j}, \tag{5.83}$$

where $\vec{s} \in \{-1, +1\}^{|\mathcal{B}|}$. $T_{\vec{s}}$ is graph-diagonal and

$$\langle\vec{k}|T_{\vec{s}}|\vec{k}\rangle = \begin{cases} 1, & \text{if } (-1)^{k_j} = s_j \text{ for all } j \in \mathcal{B} \\ 0, & \text{otherwise} \end{cases}. \tag{5.84}$$

Now, we note that

$$\langle\vec{k}|P_i|\vec{k}\rangle = \langle\vec{l}|P_i|\vec{l}\rangle, \text{ if } k_j = l_j \text{ for all } j \in \mathcal{B}. \tag{5.85}$$

5.6 Proofs

This follows from the fact that, if $|\vec{k}\rangle$ and $|\vec{l}\rangle$ have the same bit values on all qubits in $\mathcal{B}$, it is possible to obtain $|\vec{k}\rangle$ from $|\vec{l}\rangle$ by applying operators Z_j on qubits $j \notin \mathcal{B}$. Since P_i only has Z-operators (or $\mathbb{1}$) on these qubits, it commutes with Z_j, $j \notin \mathcal{B}$, and one has

$$\langle \vec{k}|P_i|\vec{k}\rangle = \langle \vec{l}|(\prod_j Z_j)P_i(\prod_j Z_j)|\vec{l}\rangle = \langle \vec{l}|P_i|\vec{l}\rangle. \tag{5.86}$$

Equation (5.85) implies that $\langle \vec{k}|\max P_i|\vec{k}\rangle$ only depends on the bit values k_j with $j \in \mathcal{B}$. We can therefore set $\alpha_{\vec{s}} = \langle \vec{k}|\max P_i|\vec{k}\rangle$, where $\vec{s} \in \{-1,+1\}^{|\mathcal{B}|}$ and $s_j = (-1)^{k_j}$ for all $j \in \mathcal{B}$. Then, we have

$$\max_{i=1,\dots,m} P_i = \sum_{\vec{s}} \alpha_{\vec{s}} T_{\vec{s}}. \tag{5.87}$$

The operators $T_{\vec{s}}$ are invariant under partial transposition, since $\mathcal{B}$ only consists of qubits that are not neighbors of each other [cf. Eq. (5.83)]. Thus, $\max_{i=1,\dots,m} P_i$ is invariant under any partial transposition. □

Let us now come to the main part of the proof of Lemma 36.

Proof. We write the given fully decomposable witnesses W_i in the form

$$W_i = \frac{1}{2}\mathbb{1} - |G\rangle\langle G| - \frac{1}{2}P_+^{(i)}, \tag{5.88}$$

where

$$P_+^{(i)} = \sum_s \prod_{j \in \mathcal{B}_i} \gamma_j^{s_j}. \tag{5.89}$$

If we introduce the shorthand notation

$$\max_i P_+^{(i)} = \sum_{\vec{k} \in \{0,1\}^n} |\vec{k}\rangle\langle \vec{k}| \max_{i=1,\dots,m} \langle \vec{k}|P_+^{(i)}|\vec{k}\rangle, \tag{5.90}$$

we can write the operator of Eq. (5.17) as

$$W = \frac{1}{2}\mathbb{1} - |G\rangle\langle G| - \frac{1}{2}\max_i P_+^{(i)}. \tag{5.91}$$

We need to prove that this is indeed a fully decomposable witness. In order to do so, we proceed in two steps. First, we prove that there is a positive operator P_M for every M which is independent from i such that $(W_i - P_M)^{T_M} \geq 0$ holds for all i. Second, we use the positive operators P_M of the first step to prove that the operator of Eq. (5.91) is a fully decomposable witness.

First step — Let us show that, for a given M, there exists a positive operator P_M independent from i that obeys

5 Entanglement witnesses for graph states

$$(W_i - P_M)^{T_M} \geq 0 \tag{5.92}$$

for all i. In order to prove this, we apply the algorithm for the construction of such operators given in the proof of Lemma 34. However, instead of applying it to any of the sets $\mathcal{B}_i$ directly, we construct a new set $\mathcal{A}$ out of these sets $\mathcal{B}_i$. Although the set $\mathcal{A}$ will contain at least as many qubits as the largest one of the sets $\mathcal{B}_i$, in most cases even more qubits, it still obeys the condition that no two qubits in $\mathcal{A}$ have a neighbor in common. Therefore, we can then apply the algorithm to it.

First, we assume that for any qubit $\beta_j^{(i)}$ from any subset $\mathcal{B}_i$, there is, in every other set $\mathcal{B}_k$ a qubit $\beta_l^{(k)}$ that has the same neighborhood as $\beta_j^{(i)}$. In principle, according to condition (ii) of Lemma 36, there can also be subsets $\mathcal{B}_k$ in which no qubit has the same neighborhood as $\beta_j^{(i)}$. However, adding the qubit $\beta_j^{(i)}$ itself to such a subset $\mathcal{B}_k$ causes the mentioned assumption to hold. After this addition, $\mathcal{B}_k$ still fulfills condition (i), since there was no qubit in $\mathcal{B}_k$ that had a neighbor in common with $\beta_j^{(i)}$ before the addition according to condition (ii).

Furthermore, for a more convenient notation, we relabel the qubits $\beta_j^{(i)}$ in such a way that the two qubits $\beta_j^{(i)} \in \mathcal{B}_i$ and $\beta_j^{(k)} \in \mathcal{B}_k$ with the same subscript j also have the same neighborhood. According to the assumption in the last paragraph, there is exactly one qubit in $\mathcal{B}_k$ that has the same neighborhood as $\beta_j^{(i)}$.

Before constructing $\mathcal{A}$ and applying the algorithm, we perform the aforementioned transformation $G \to G'$ (cf. Sec. 5.6) for the given partition M by deleting all edges that connect qubits in the same partition. Note that two qubits that had the same neighborhood in G do not need to have the same neighborhood in G' anymore.

As argued after Eq. (5.51), this transformation changes Eq. (5.92) into

$$\langle \vec{k}| (W_i' - P_M')^{T_M} |\vec{k}\rangle \geq 0, \tag{5.93}$$

which has to be shown for all vectors $|\vec{k}\rangle$ of the basis given by the transformed generators $g_i' = (A \otimes B) g_i (A^\dagger \otimes B^\dagger)$, for all i and for all M. Here, $W_i' = (A \otimes B) W_i (A^\dagger \otimes B^\dagger)$ and $P_M' = (A \otimes B) P_M (A^\dagger \otimes B^\dagger)$, where $A = \prod_{(j,l)} C_{j,l}$ and $B = \prod_{(j,l)} C_{j,l}$ are the unitary operators that correspond to the deletion of the edges in M and in $\overline{M}$, respectively. Therefore, A acts on qubits in M and B on qubits in $\overline{M}$ (which is not obvious from our above notation). Also, we have used that W_i' and P_M' are diagonal in the basis given by the vectors $|\vec{k}\rangle$.

Then, we construct a set $\mathcal{A} = \{\alpha_i\}$ in the following way.

1. Start with the empty set $\mathcal{A} = \{\}$.

2. Let $j = 1$.

3. If all qubits $\beta_j^{(i)}$, $i = 1, \ldots, m$, from the M subsets $\mathcal{B}_i$ are in the same partition, then add $\beta_j^{(1)}$ to the set $\mathcal{A}$. Otherwise, there exists a qubit $\beta_j^{(x)}$ that is in the opposite partition as $\beta_j^{(1)}$. Then, add both $\beta_j^{(x)}$ and $\beta_j^{(1)}$ to $\mathcal{A}$.

5.6 Proofs

4. Increase j by one. If $j \leq |\mathcal{B}_1|$, repeat the last step. Otherwise, the construction is finished. Note that any other set $\mathcal{B}_i$ contains the same number of qubits as $\mathcal{B}_1$.

Step 3 is the crucial one and we note the following points: If all qubits $\beta_j^{(i)}$, $i = 1, \ldots, m$, are in the same partition, we add $\beta_j^{(1)}$ from $\mathcal{B}_1$ to $\mathcal{A}$. In principle, in this case one can instead add the jth qubit $\beta_j^{(i)}$ from any other set $\mathcal{B}_i$ to $\mathcal{A}$, since all of them have the same neighborhood even after the transformation $G \to G'$, as they are all in the same partition.

In the other case, there are two qubits $\beta_j^{(1)}$ and $\beta_j^{(x)}$ in opposite partitions. Then, both of them are added to $\mathcal{A}$. However, since they are in opposite partitions and had the same neighborhood in graph G, they cannot have a neighbor in common after the transformation $G \to G'$. Such a neighbor in common would have to be in the opposite partition as $\beta_j^{(1)}$, in order to be its neighbor in graph G', but at the same time in the opposite partition as $\beta_j^{(x)}$. This is impossible since $\beta_j^{(1)}$ and $\beta_j^{(x)}$ are in opposite partitions.

Together with the fact that any two qubits of the set $\mathcal{B}_1$ do not have a neighbor in common according to the conditions of Lemma 36, this shows that no two qubits in $\mathcal{A}$ have a neighbor in common (in graph G'). Also, there cannot be two qubits in $\mathcal{A}$ which are neighbors of each other, since these must have also been neighbors in G, which contradicts the conditions of Lemma 36.

Now, we use the algorithm presented after Eq. (5.57) to construct P'_M, but we apply it the qubits α_i in the set $\mathcal{A}$ instead of the qubits in set $\mathcal{B}$ as in the original algorithm. Again, P'_M is a sum over projectors onto graph basis states [cf. Eq. (5.60)].

Since $P_+^{(i)}$ is invariant under partial transposition, Eq. (5.52) implies that $P_+^{\prime(i)} = (A \otimes B) P_+^{(i)} (A^\dagger \otimes B^\dagger)$ is also invariant. Using the explicit form of W_i', we can thus rewrite Eq. (5.93) as

$$\frac{1}{2} - \frac{1}{2} \langle \vec{k} | P_+^{\prime(i)} | \vec{k} \rangle - \langle \vec{k} | (|G'\rangle\langle G'| + P'_M)^{T_M} | \vec{k} \rangle \geq 0 . \quad (5.94)$$

For a given i and M, we consider two cases for the vectors $|\vec{k}\rangle$. Then, the reasoning is analogous to the two cases in Sec. 5.6.

1. $\boxed{\langle \vec{k} | P_+^{\prime(i)} | \vec{k} \rangle = 1}$

Since $P_+^{\prime(i)}$ has the form of Eq. (5.89), but with the transformed generators g_i', and is therefore a sum of projectors as in Eq. (5.60), in this case there are two qubits $j, l \in \mathcal{B}_i$ with $k_j = k_l = 1$. Moreover, Eq. (5.94) reduces to

$$-\langle \vec{k} | (|G'\rangle\langle G'| + P'_M)^{T_M} | \vec{k} \rangle \geq 0 . \quad (5.95)$$

Per construction, $\mathcal{A}$ contains qubits j and l or qubits that have the same neighborhood as qubits j and l. Therefore, the algorithm constructs an operator P'_M that contains only projectors $|\vec{a}\rangle\langle\vec{a}|$ that obey $a_j = a_l = 0$. This can be seen in step 3 of the algorithm for the construction of P'_M,

5 Entanglement witnesses for graph states

in which only qubits in the neighborhood of qubits in $\mathcal{A}$ are flipped. Note that $|G'\rangle = |0\ldots 0\rangle$ and therefore also here, the jth and the lth bit equal zero. For this reason, Lemma 41 implies that

$$-\langle \vec{k}|(|G'\rangle\langle G'| + P'_M)^{T_M}|\vec{k}\rangle = 0 \tag{5.96}$$

and that Eqs. (5.95) and (5.94) hold.

2. $\boxed{\langle \vec{k}|P'^{(i)}_+|\vec{k}\rangle = 0}$

If $P'_M \neq 0$, it has 2^{r-1} terms, where r is the number of qubits α_i that have a neighbor in graph G'. Since no two qubits in $\mathcal{A}$ have a neighbor in common, one can invoke Lemmata 42 and 44 to show that Eq. (5.94) holds for $|\vec{k}\rangle$ with $\langle \vec{k}|P'^{(i)}_+|\vec{k}\rangle = 0$.

In the case $P_M = 0$, there must be at least one pair of qubits in G' which are connected with each other and in opposite partitions. These can be transformed into a Bell pair via LOCC, such that Eq. (5.94) holds.

Thus, Eq. (5.92) holds for all i and the constructed operators $P_M = \left(A^\dagger \otimes B^\dagger\right) P'_M \left(A \otimes B\right)$.

Second step — In the second step, we can now use the positive operators P_M constructed in the last step to show that the operator W of Eq. (5.91) is a fully decomposable witness. In order to do so, we show that, for every M, the positive semidefinite operator P_M of the last step fulfills $(W - P_M)^{T_M} \geq 0$. Since W and the operators P_M are graph-diagonal, it is enough to show the positivity of $\langle \vec{k}|(W - P_M)^{T_M}|\vec{k}\rangle$ for all $|\vec{k}\rangle$.

We define

$$R_i = \max_j P^{(j)}_+ - P^{(i)}_+. \tag{5.97}$$

Note that R_i is invariant under partial transposition, as $P^{(i)}_+$ does not contain generators of neighboring qubits, and is therefore invariant, and $\max_j P^{(j)}_+$ is invariant according to Lemma 45. Moreover, for a given $|\vec{k}\rangle$, let i_0 be the value of i that maximizes $\langle \vec{k}|P^{(i)}_+|\vec{k}\rangle$. Then, we have

$$\langle \vec{k}|(W - P_M)^{T_M}|\vec{k}\rangle$$
$$= \langle \vec{k}|\left(W_{i_0} - \frac{1}{2}R_{i_0} - P_M\right)^{T_M}|\vec{k}\rangle$$
$$= \langle \vec{k}|(W_{i_0} - P_M)^{T_M}|\vec{k}\rangle - \frac{1}{2}\langle \vec{k}|R_{i_0}|\vec{k}\rangle$$
$$\geq 0. \tag{5.98}$$

5.6 Proofs

In the first line, we have employed the definitions in Eqs. (5.88), (5.91) and (5.97). In the second line, we have used the invariance of R_i under partial transposition. Finally, for the positivity, we used Eq. (5.92) and

$$\begin{aligned}\langle \vec{k}|R_{i_0}|\vec{k}\rangle &= \langle \vec{k}|\max_j P_+^{(j)}|\vec{k}\rangle - \langle \vec{k}|P_+^{(i_0)}|\vec{k}\rangle \\ &= \langle \vec{k}|P_+^{(i_0)}|\vec{k}\rangle - \langle \vec{k}|P_+^{(i_0)}|\vec{k}\rangle \\ &= 0\,.\end{aligned} \quad (5.99)$$

Thus, the operator W of Eq. (5.17) is a fully decomposable witness. □

Fully PPT witnesses for arbitrary graph states (Lemma 37)

Proof. The proof that we present here is similar to the proof of Lemma 34 (cf. Sec. 5.6). In fact, it is much shorter, since we do not have to provide a construction algorithm for the positive operators P_M, as these equal zero for fully PPT witnesses.

Here, we have to prove that
$$W_G^{T_M} \geq 0 \qquad (5.100)$$

holds for every strict subset M of the set of all qubits [cf. Eq. (5.49) in the proof of Lemma 34]. Since $P_+ = \sum_{\vec{s}} \prod_{i \in \mathcal{B}} \gamma_i^{s_i}$ is invariant under partial transposition and $W_G^{T_M}$ is graph-diagonal, one can plug Eq. (5.20) into Eq. (5.100) to obtain

$$\frac{1}{2} - \langle \vec{k}|\,(|G\rangle\langle G|)^{T_M}\,|\vec{k}\rangle - \left(\frac{1}{2} - \frac{1}{2^{m(\vec{k})}}\right)\langle \vec{k}|P_+|\vec{k}\rangle \geq 0 \qquad (5.101)$$

which has to hold for all M and all graph state basis vectors $|\vec{k}\rangle$ [cf. Eq. (5.69)]. Here, $m(\vec{k})$ denotes the number of ones in the binary vector $\vec{k}$ that are on qubits contained in $\mathcal{B}$. These correspond to -1s in a sign vector $\vec{s}$ (and zeros in $\vec{k}$ correspond to $+1$s in $\vec{s}$). In formulas, $m(\vec{k}) = \left(b - \sum_{i \in \mathcal{B}}(-1)^{k_i}\right)/2$.

As before [cf. Eq. (5.53)], we transform graph G into the graph G' by deleting all edges that connect qubits in the same partition. As in the proof of Lemma 34, we now distinguish two cases.

1. $\boxed{\langle \vec{k}|P'_+|\vec{k}\rangle = 0}$

   In this case, Eq. (5.101) can be rewritten as

   $$\langle \vec{k}|\,(|G'\rangle\langle G'|)^{T_M}\,|\vec{k}\rangle \leq \frac{1}{2}\,. \qquad (5.102)$$

   This equation holds, as we have already argued after Eq. (5.78), since there must at least be two neighboring qubits in opposite partitions.

2. $\boxed{\langle\vec{k}|P'_+|\vec{k}\rangle \neq 0 \Leftrightarrow \langle\vec{k}|P'_+|\vec{k}\rangle = 1}$

Here, we need to prove that

$$\langle\vec{k}|\,(|G'\rangle\langle G'|)^{T_M}\,|\vec{k}\rangle \leq \frac{1}{2^{m(\vec{k})}}\,, \tag{5.103}$$

where $m(\vec{k})$ is the number of ones in $\vec{k}$ on qubits in $\mathcal{B}$. If there is a qubit $i \in \mathcal{B}$ with $k_i = 1$ with only neighbors that are in the same partition as qubit i, then Lemma 41 applies and the left-hand side of Eq. (5.103) vanishes.

In the case in which no qubit $i \in \mathcal{B}$ with $k_i = 1$ has only neighbors in the same partition, Lemmata 42 and 44 imply that

$$\langle\vec{k}|\,(|G'\rangle\langle G'|)^{T_M}\,|\vec{k}\rangle \leq \frac{1}{2^b}\,, \tag{5.104}$$

where $b = |\mathcal{B}|$. Since $m(\vec{k}) \leq b$, Eq. (5.103) holds.

□

Extended construction of fully PPT witnesses (Lemma 38)

Proof. Here, we prove that the operator W of Eq. (5.22) is a fully PPT witness. To this end, we write the given fully PPT witnesses W_i as

$$W_i = \frac{1}{2}\mathbb{1} - |G\rangle\langle G| - P_+^{(i)}\,, \tag{5.105}$$

with the definition

$$P_+^{(i)} = \sum_{\vec{s}} \left(\frac{1}{2} - \frac{1}{2^{m(\vec{s})}}\right) \prod_{j \in \mathcal{B}_i} \gamma_j^{s_j}\,. \tag{5.106}$$

Here, $m(\vec{s})$ is the number of elements $s_j = -1$ in $\vec{s}$, i.e., $m(\vec{s}) = \left(|\mathcal{B}_i| - \sum_{j=1}^{|\mathcal{B}_i|} s_j\right)/2$. As we did before, we now introduce the shorthand notation

$$\max_i P_+^{(i)} = \sum_{\vec{k}} |\vec{k}\rangle\langle\vec{k}| \max_{i=1,\ldots,m} \langle\vec{k}|P_+^{(i)}|\vec{k}\rangle\,, \tag{5.107}$$

and can thus write the operator of Eq. (5.22) as

$$W = \frac{1}{2}\mathbb{1} - |G\rangle\langle G| - \max_i P_+^{(i)}\,. \tag{5.108}$$

5.6 Proofs

Now, we proceed similarly to the proof of Lemma 36 (Sec. 5.6), but, since fully PPT witnesses have $P_M = 0$, we do not need to construct such operators here. Therefore, the proof in this section is much shorter.

Again, we define
$$R_i = \max_j P_+^{(j)} - P_+^{(i)}, \tag{5.109}$$

which is invariant under any partial transposition due to Lemma 45. For a given $|\vec{k}\rangle$, let i_0 be the value of i that maximizes $\langle \vec{k}|P_+^{(i)}|\vec{k}\rangle$. Then, we have

$$\begin{aligned}
\langle \vec{k}|W^{T_M}|\vec{k}\rangle &= \langle \vec{k}|\left(W_{i_0} - R_{i_0}\right)^{T_M}|\vec{k}\rangle \\
&= \langle \vec{k}|W_{i_0}^{T_M}|\vec{k}\rangle - \langle \vec{k}|R_{i_0}|\vec{k}\rangle \\
&\geq 0.
\end{aligned} \tag{5.110}$$

In the first line, we plugged in the definitions in Eqs. (5.105), (5.108) and (5.109). In the next step, we used that R_i is invariant under partial transposition. Finally, for the positivity, we used that the operators W_i are fully PPT witnesses and that

$$\begin{aligned}
\langle \vec{k}|R_{i_0}|\vec{k}\rangle &= \langle \vec{k}|\max_j P_+^{(j)}|\vec{k}\rangle - \langle \vec{k}|P_+^{(i_0)}|\vec{k}\rangle \\
&= \langle \vec{k}|P_+^{(i_0)}|\vec{k}\rangle - \langle \vec{k}|P_+^{(i_0)}|\vec{k}\rangle \\
&= 0.
\end{aligned} \tag{5.111}$$

Thus, the operator W of Eq. (5.22) is a fully PPT witness.

□

Fully PPT witness for the 2D cluster state (Lemma 39)

Proof. In order to show that the operator of Eq. (5.32) is a fully PPT witness, we provide two lemmata first. The first one specifies some conditions, under which the overlap of the partially transposed 2D cluster state with another basis vector vanishes for certain bipartitions. The second lemma provides an upper bound for the largest Schmidt coefficient of the 2D cluster state for bipartitions in which no partition contains less than two qubits. Note that both lemmata hold for 2D cluster states of $n \times n$ qubits with $n > 2$ and it is therefore straightforward to see that the proof presented here also holds for more than 16 qubits.

Lemma 46. *Given a 2D cluster $|\text{Cl}_{n \times n}\rangle$ of n^2 qubits. Consider an arbitrary qubit q of these. Let $|\vec{a}\rangle$ be a state of the corresponding graph state basis. If there is a qubit $i \neq q$ with $a_i = 1$ and there is a* qubit $j \in \mathcal{N}(q)$ with $a_j = 0$, then

$$\langle \vec{a} | \left(|\mathrm{Cl}_{\mathrm{n}\times\mathrm{n}}\rangle\langle \mathrm{Cl}_{\mathrm{n}\times\mathrm{n}}| \right)^{T_q} | \vec{a} \rangle = 0 \, . \tag{5.112}$$

Proof. First, note that $|\mathrm{Cl}_{\mathrm{n}\times\mathrm{n}}\rangle$ can be written as $|0\ldots 0\rangle$ in its graph basis. Thus, according to Lemma 41, Eq. (5.112) holds if $i \notin \mathcal{N}(q)$, independent of the condition on qubit j.
Thus, it remains to show Eq. (5.112) for the case $i \in \mathcal{N}(q)$. Due to Eq. (2.69), we can write Eq. (5.112) as

$$\langle \vec{a} | \left(|0\ldots 0\rangle\langle 0\ldots 0| \right)^{T_q} | \vec{a} \rangle$$
$$= \mathrm{Tr}\left\{ \left[\prod_l \frac{1}{2}(\mathbb{1} + g_l) \right]^{T_q} \prod_k \frac{1}{2}[\mathbb{1} + (-1)^{a_k} g_k] \right\} \, . \tag{5.113}$$

To simplify this expression, we note that

$$\prod_k \frac{1}{2}[\mathbb{1} + (-1)^{a_k} g_k] = \sum_{\vec{x}} (-1)^{\vec{a}\vec{x}} \prod_k g_k^{x_k} \, , \tag{5.114}$$

where the sum runs over all binary vectors $\vec{x}$ of length n^2.

Moreover, we define a boolean function f that characterizes the action of the partial transposition on products of generators in the following way:

$$f : \{0, 1\}^{n^2} \to \{0, 1\} \tag{5.115}$$

$$\vec{x} \mapsto f(\vec{x}) = \begin{cases} 0, & \text{if } \left(\prod_{i=1}^n g_i^{x_i} \right)^{T_q} = \prod_{i=1}^n g_i^{x_i} \\ 1, & \text{if } \left(\prod_{i=1}^n g_i^{x_i} \right)^{T_q} = -\prod_{i=1}^n g_i^{x_i} \end{cases} \, .$$

Note that f depends on q. With these definitions, we can write

$$\left(\prod_l g_l^{y_l} \right)^{T_q} = (-1)^{f(\vec{y})} \prod_l g_l^{y_l} \, . \tag{5.116}$$

Applying Eqs. (5.114) and (5.116) to simplify Eq. (5.113) results in

5.6 Proofs

$$\langle \vec{a} | (|0\ldots 0\rangle\langle 0\ldots 0|)^{T_q} |\vec{a}\rangle$$

$$= 4^{-n^2} \text{Tr} \left\{ \left[\sum_{\vec{y}} (-1)^{f(\vec{y})} \left(\prod_k g_k^{y_k} \right) \right] \left[\sum_{\vec{x}} (-1)^{\vec{a}\vec{x}} \prod_l g_l^{x_l} \right] \right\}$$

$$= 2^{-n^2} \sum_{\vec{x} \in \{0,1\}^{n^2}} (-1)^{\vec{a}\vec{x} + f(\vec{x})} \,. \tag{5.117}$$

In the last step, we have used $\text{Tr}(\prod_k g^{y_k} \prod_l g^{x_l}) = 2^n \delta_{\vec{x}, \vec{y}}$ vanishes if $\vec{x} \neq \vec{y}$. Since both generators g_i and g_j have Z operator on qubit q, their product $g_i g_j$ acts trivially on qubit q. Therefore,

$$f(x_1, \ldots, x_i, \ldots, x_j, \ldots, x_{n^2})$$
$$= f(x_1, \ldots, x_i \oplus 1, \ldots, x_j \oplus 1, \ldots, x_{n^2}) \,. \tag{5.118}$$

Furthermore, since $a_i = 1$, $a_j = 0$, a term in the sum of Eq. (5.117) with $x_i = 1$, $x_j = 1$ will have the opposite as the same term with x_i and x_j flipped. Also, flipping $x_i = 1$, $x_j = 0$ to $x_i = 0$, $x_j = 1$ changes the sign of the corresponding term. Thus, the sum in Eq. (5.117) vanishes. □

Lemma 47. *Given a 2D cluster state $|Cl_{n \times n}\rangle$ with periodic boundary conditions, $n > 2$ and a bipartition $M|\overline{M}$. Let λ_i be the Schmidt coefficients of $|Cl_{n \times n}\rangle$ with respect to this bipartition. If $|M| \geq 2$ and $|\overline{M}| \geq 2$, then*

$$\max_i \lambda_i^2 \leq \frac{1}{4} \tag{5.119}$$

Proof. To prove this claim, we provide an LOCC protocol for every possible case which results in two disconnected Bell pairs. Since LOCC does not decrease the largest Schmidt coefficient and a single Bell pair has a Schmidt coefficient of $1/\sqrt{2}$, the upper bound of Eq. (5.119) follows from it.

Due to the assumptions, there are at least two qubits $i, k \in M$ which each must have a neighbor in $\overline{M}$, say $j \in \mathcal{N}(i), l \in \mathcal{N}(k)$ and $j \neq l$. Let us now describe how one can create a Bell pair between i and j and one between k and l, both of which are disconnected from the rest of the graph.

If i and k can be chosen in such a way that the qubits i, j, k, l are not connected with each other except for the two edges between i and j, k and l, measuring out all qubits besides i, j, k and l results in the desired two Bell pairs.

Now, consider the case that the qubits i, j, k, l have more connections amongst each other than the two connections that will be used for the Bell pairs. Then, we first delete all edges that connect qubits of the same partition, which is an LU operation. Then, there are four possible situations as shown in Fig. 5.6. Note that edges that connect the four qubits with other qubits are drawn dashed and in gray, since they might have been deleted by the last operation (and are not needed for the protocol anyway). Moreover, situations a) and b) are equivalent to a number of other ones that we did not explicitly draw, in which all the qubits i, j, k, l form a one-dimensional chain [and not a square as in c) and d)]. Note that, if i and k are disconnected in d), j and l must also be disconnected, as i and k being in the same partition implies that also j and l are in the same one.

In cases a) and d), simply measuring out all qubits besides i, j, k, l results in the two Bell pairs.

For cases b) and c), the procedure is slightly more complicated. In case b), we create an edge between the qubits j and l, which are in the same partition, via an LU operation. Then, local complementation on l deletes the unwanted edge between j and k. Afterwards, we can delete the edge between j and l again. Note that the local complementation possibly also creates (or deletes) other edges between neighbors of l. Note that, however, it does not delete the important edges between i and j, k and l. Moreover, our last step is to measure all qubits besides i, j, k, l which also deletes any such edges that might have been created.

Finally, consider case c). Here, it is not enough to simply consider the qubits i, j, k, l, since the four-qubit ring cluster that they build (disregarding any connections to other qubits) is LU-equivalent with a single Bell pair. However, a closer look at situation c) shows that it actually implies that there are four qubits as in b) or in d).

Assume that there is any qubit that neighbors any of the qubits i, j, k or l — say k — and is in the opposite partition as k. Note that this qubit could not be, in the case of a 3×3-cluster, a neighbor of j or l, since these qubits also lie in the opposite partition as k (and edges between qubits that are in the same partition have been deleted). Thus, we have a situation as in b).

Assume now that there is no neighbor of any of the qubits i, j, k, l that is in the opposite partition as the qubit it neighbors. In other words, each of the four qubits has only neighbors in the same partition. Then, there is another pair of qubits which is in opposite partitions, namely a neighbor of k and one of l. This means that there we have a situation as in d).

Thus, we always obtain two Bell pairs and the proof is finished.

□

Now, we return to the proof of Lemma 39 which is easy to prove having the last two lemmata in mind. We consider a 2D cluster state of $n \times n$ qubits with $n \geq 3$. As in the proofs before, we write the witness in the form

$$W_{n \times n} = \frac{1}{2}\mathbb{1} - |Cl_{n\times n}\rangle\langle Cl_{n\times n}| - \frac{1}{4}P_+ \,, \qquad (5.120)$$

where we defined $P_+ = \sum_{\vec{k}} |\vec{k}\rangle\langle\vec{k}| \max_{(i,j)} \langle\vec{k}|D_{(i,j)}|\vec{k}\rangle$ with the operators $D_{(i,j)}$ of Eq. (5.31).

Every generator in P_+ is neighbored by either two or no other generator. Therefore, P_+ does not have a Y on any qubit in M. Thus, P_+ is invariant under any partial transposition. Since the witness of Eq. (5.120) is diagonal in the graph state basis, we need to prove that

$$\frac{1}{2} - \langle\vec{k}|\left(|Cl_{n\times n}\rangle\langle Cl_{n\times n}|\right)^{T_M}|\vec{k}\rangle - \frac{1}{4}\langle\vec{k}|P_+|\vec{k}\rangle \geq 0 \qquad (5.121)$$

5.6 Proofs

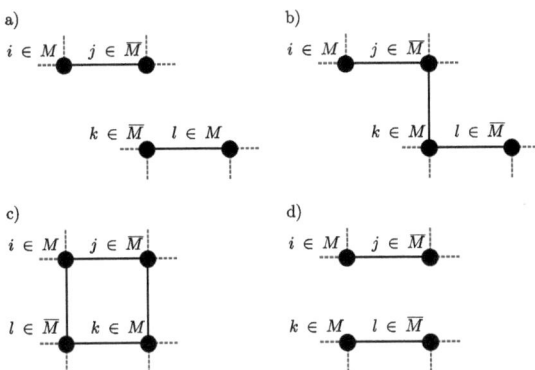

Figure 5.6: In a 2D cluster state and for any bipartition $M|\overline{M}$, where there are at least two qubits in each partition, one can always obtain two Bell pairs via LOCC operations. Here, we illustrate all possible cases, in which the two Bell pairs are connected to each other in the original 2D cluster state. Note that edges between qubits in the same partition have been deleted. Also, edges that lead to qubits which are not part of the Bell pairs are shown as dashed lines.

holds for all partitions M and all graph basis vectors $|\vec{k}\rangle$.

In the case of a vector $|\vec{k}\rangle$ with $\langle\vec{k}|P_+|\vec{k}\rangle = 0$, we need to show, according to Lemma 42, that the largest Schmidt coefficient of $|\text{Cl}_{n\times n}\rangle$ with respect to $M|\overline{M}$ is smaller than (or equal to) $1/\sqrt{2}$. This is trivial, since every connected graph state can be distilled to at least one Bell pair via LOCC operations.

In the case of a vector $|\vec{k}\rangle$ with $\langle\vec{k}|P_+|\vec{k}\rangle = 1$, we have to prove that

$$\langle\vec{k}|\left(|\text{Cl}_{n\times n}\rangle\langle\text{Cl}_{n\times n}|\right)^{T_M}|\vec{k}\rangle \leq \frac{1}{4}. \tag{5.122}$$

If $|M| \geq 2$ and $|\overline{M}| \geq 2$, we can apply Lemma 47 (and Lemma 42) to show this.

Thus, it remains to show Eq. (5.122) for the case that $|M| = 1$ or $|\overline{M}| = 1$. Since $W^{T_M} \geq 0 \Leftrightarrow W^{T_{\overline{M}}} \geq 0$, we can assume w.l.o.g. that $|M| = 1$ and write $M = \{q\}$. Moreover, $\langle\vec{k}|P_+|\vec{k}\rangle = 1$ together with the form of P_+ [cf. Eq. (5.32)] implies that there are two diagonals which we denote by $\mathcal{D}^{(x)}$ and $\mathcal{D}^{(y)}$ here (cf. Fig. 5.7), on which $|\vec{k}\rangle$ has an odd number of ones. In formulas,

5 Entanglement witnesses for graph states

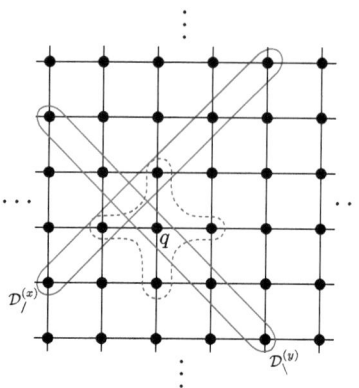

Figure 5.7: The proof of the fully PPT witness for an $n \times n$ 2D cluster state considers a one-particle partition $M = \{q\}$ and distinguishes different cases as depicted here. The diagonals $\mathcal{D}_/$ and $\mathcal{D}_\backslash$ mentioned in the text are marked. For more details, see text.

$$\prod_{l \in \mathcal{D}_/^{(x)}} g_l |\vec{k}\rangle = -|\vec{k}\rangle, \qquad (5.123)$$

$$\prod_{l \in \mathcal{D}_\backslash^{(y)}} g_l |\vec{k}\rangle = -|\vec{k}\rangle . \qquad (5.124)$$

Note that Lemma 46 can be used to show that Eq. (5.122) holds if there exists another qubit $i \neq q$ with $k_i = 1$ and a qubit j in the neighborhood of q with $k_j = 0$.

Since it is impossible that all qubits $i \neq q$ are zero as this would contradict Eqs. (5.123), (5.124) and the fact that $\mathcal{D}_/^{(x)}$ and $\mathcal{D}_\backslash^{(y)}$ have non qubit in common, the only other case is that all qubits in the neighborhood of q equal one.

In this case, we only need to consider the five qubits in $\widetilde{\mathcal{N}}(q) = \mathcal{N}(q) \cup q$ (marked by a dashed line in Fig. 5.7). Note that the following argumentation is independent from the value of k_q itself. Since $\mathcal{D}_/^{(x)}$ and $\mathcal{D}_\backslash^{(y)}$ have no qubit in common, it is impossible to choose q in such a way that both the intersection of $\widetilde{\mathcal{N}}(q)$ with $\mathcal{D}_/^{(x)}$ and the intersection of $\widetilde{\mathcal{N}}(q)$ with $\mathcal{D}_\backslash^{(y)}$ consist of an odd number of qubits. One of the two intersections always has two or zero qubits. Without loss of generality, we assume that the intersection of $\widetilde{\mathcal{N}}(q)$ with $\mathcal{D}_/^{(x)}$ has an even number of qubits. An example for this situation is given in Fig. 5.7. Then, due to Eq. (5.123), there is a qubit in $\mathcal{D}_/^{(x)}$ which equals one and to which therefore Lemma 46 can be applied.

5.6 Proofs

This shows that Eq. (5.122) holds in all cases and finishes the proof. □

Values of the entanglement monotone for graph states (Lemma 40)

Proof. (Lemma 40) — We define the set of all appropriately normalized witnesses that are decomposable with respect to bipartition $M|\overline{M}$ as

$$\mathcal{W}_M = \{W | \exists\, P, Q \text{ such that } 0 \leq P, Q \leq \mathbb{1} \text{ and } W = P + Q^{T_M}\}, \tag{5.125}$$

such that the set of all similarly normalized, fully decomposable witnesses $\mathcal{W}$ of Eq. (4.5) obeys $\mathcal{W} = \cap_M \mathcal{W}_M$. Since $\mathcal{W} \subseteq \mathcal{W}_{M_0}$ for any fixed bipartition $M_0|\overline{M_0}$, we have

$$\mathcal{E}(\varrho) \leq - \min_{W \in \mathcal{W}_{M_0}} \mathrm{Tr}(W\varrho). \tag{5.126}$$

According to Lemma 29, $\min_{W \in \mathcal{W}_{M_0}} \mathrm{Tr}(W\varrho)$ equals the negativity with respect to the bipartition $M_0|\overline{M_0}$. If all particles are qubits, we now choose any bipartition $M_0|\overline{M_0}$ in which M_0 only contains one particle, e.g. the bipartition $A|BCD\ldots$. Then,

$$N(\varrho) \leq - \min_{W \in \mathcal{W}_{M_0}} \mathrm{Tr}(W\varrho)$$

$$\leq \max_{|\varphi\rangle} \left(- \min_{W \in \mathcal{W}_{M_0}} \mathrm{Tr}(W|\varphi\rangle\langle\varphi|) \right)$$

$$= \frac{1}{2}. \tag{5.127}$$

Here, we use that the expectation value is a linear function and must therefore attain its maximum on a pure state $|\varphi\rangle$. Moreover, the last equality stems from the fact that the negativity with respect to a bipartition $A|BCD\ldots$, where A is a single qubit, can maximally take on the value one half. This maximum is obtained for the Bell state $|\psi^+\rangle = (|00\rangle + |11\rangle)/\sqrt{2}$.

If not all particles are qubits, one chooses M_0 to consist of a particle that has the smallest dimension of all occurring particles. For example, if A and B are four-level particles and C and D are qutrits, then $M_0 = C$ is a valid choice. In this case, the value of one half in the last line of Eq. (5.127) must be replaced by $(d_{\min} - 1)/2$, where $d_{\min}$ is the dimension of the particle with lowest dimension. This maximum is obtained for the state $|\psi\rangle = \sum_{i=0}^{d-1} |ii\rangle^{\otimes n}/\sqrt{d_{\min}}$. It is the maximal value for the negativity with respect to the given bipartition as can be easily seen using the Schmidt decomposition and the fact that $|\psi\rangle\langle\psi|^{T_{M_0}}$ has $d_{\min}(d_{\min} - 1)/2$ negative eigenvalues that all equal $-1/d_{\min}$.

We now know that the entanglement measure is upper-bounded by one half for states that consist only of qubits. Let us now show that, for graph states, the lower bound is also one half. This is easy to see, since we only have to pick one witness $W_G \in \mathcal{W}$ for the given graph state $|G\rangle\langle G|$. Such a witness is the projector witness

$$W_G = \frac{1}{2}\mathbb{1} - |G\rangle\langle G| \qquad (5.128)$$

which is even a fully PPT witness. It remains to show that this witness also obeys

$$\mathbb{1} \geq W_G^{T_M} = \frac{1}{2}\mathbb{1} - (|G\rangle\langle G|)^{T_M} \geq 0 \,. \qquad (5.129)$$

Here, positivity follows from the projector witness being a fully PPT witness. For the inequality on the left, we need to show that

$$\langle \vec{k}| (|G\rangle\langle G|)^{T_M} |\vec{k}\rangle \geq -\frac{1}{2} \qquad (5.130)$$

holds for all $|\vec{k}\rangle$, since the partial transpose of $|G\rangle\langle G|$ is again graph-diagonal.

In order to prove Eq. (5.130), we use the Schmidt decomposition $|G\rangle = \sum_{i=1} \lambda_i |\mu_i\rangle \otimes |\nu_i\rangle$ with respect to bipartition $M|\overline{M}$ with positive and real Schmidt coefficients λ_i. Performing the partial transpose in the basis $|\mu_i\rangle \otimes |\nu_j\rangle$ allows to derive a lower bound on $\langle \vec{k}| (|G\rangle\langle G|)^{T_M} |\vec{k}\rangle$ in terms of the Schmidt coefficients in the following way:

$$\begin{aligned}
\langle \vec{k}| (|G\rangle\langle G|)^{T_M} |\vec{k}\rangle &\geq \min_{i \neq j}(-\lambda_i \lambda_j) \\
&\geq \min_i(-\lambda_i \sqrt{1-\lambda_i^2}) \\
&\geq -\frac{1}{2}\,.
\end{aligned} \qquad (5.131)$$

In the second line, we used that, as an entangled state, $|G\rangle$ has at least two non-zero Schmidt coefficients and that the squares of all coefficients must sum up to one. The last line follows from the fact that $0 < \lambda_i < 1$.

Consequently, Eq. (5.129) holds and W_G lies in $\mathcal{W}$. Therefore,

$$N(|G\rangle\langle G|) \geq -\mathrm{Tr}(W_G |G\rangle\langle G|) = \frac{1}{2}\,. \qquad (5.132)$$

Therefore, when considering the entanglement measure of Eq. (4.4), the connected graph states are the maximally entangled states. For them, the measure equals one half. □

5.7 Witnesses

Note that all witnesses are presented in their graph state basis. As before, we defined $\gamma_i^\pm = \frac{\mathbb{1} \pm g_i}{2}$. Since all witnesses are diagonal in the graph basis, we use the shorter notation $|\vec{k}\rangle\langle\cdot| = |\vec{k}\rangle\langle\vec{k}|$. Moreover, for some states one can make use of their translational symmetry. In these cases, $\mathcal{T}(\vec{k})$ denotes all translations of the bit string $\vec{k} = k_1 \ldots k_n$. For example,

$$|k_1 k_2 \mathcal{T}(k_3 k_4 k_5 k_6)\rangle\langle\cdot| = |k_1 k_2 k_3 k_4 k_5 k_6\rangle\langle\cdot| + |k_1 k_2 k_6 k_3 k_4 k_5\rangle\langle\cdot|$$
$$+ |k_1 k_2 k_5 k_6 k_3 k_4\rangle\langle\cdot| + |k_1 k_2 k_4 k_5 k_6 k_3\rangle\langle\cdot| \qquad (5.133)$$

No. 1, Bell state
$$W = \frac{\mathbb{1}}{2} - |G\rangle\langle G|$$

No. 2, GHZ$_3$
$$W = \frac{\mathbb{1}}{2} - |G\rangle\langle G|$$

No. 3, GHZ$_4$
$$W = \frac{\mathbb{1}}{2} - |G\rangle\langle G|$$

No. 4, Cl$_4$
$$W = \frac{\mathbb{1}}{2} - |G\rangle\langle G| - \frac{1}{2}\gamma_1^- \gamma_4^-$$

No. 5, GHZ$_5$
$$W = \frac{\mathbb{1}}{2} - |G\rangle\langle G|$$

No. 6, Y$_5$
$$W = \frac{\mathbb{1}}{2} - |G\rangle\langle G| - \frac{1}{2}\gamma_1^- \gamma_4^- - \frac{1}{2}\gamma_1^+ \gamma_4^- \gamma_5^-$$

No. 7, Cl$_5$
$$W = \frac{\mathbb{1}}{2} - |G\rangle\langle G| - \frac{1}{2}\gamma_1^- \gamma_5^- - \frac{1}{4}\gamma_1^+ \gamma_2^- \gamma_5^- - \frac{1}{4}\gamma_1^- \gamma_4^- \gamma_5^+$$

No. 8, R$_5$
$$W = 3\Big[- |G\rangle\langle G| + |\mathcal{T}(00001)\rangle\langle\cdot| + |\mathcal{T}(00101)\rangle\langle\cdot| + |\mathcal{T}(00111)\rangle\langle\cdot| \Big]$$
$$- |11111\rangle\langle\cdot| + |\mathcal{T}(11110)\rangle\langle\cdot| + |\mathcal{T}(11010)\rangle\langle\cdot| + |\mathcal{T}(11000)\rangle\langle\cdot|$$

No. 9, GHZ$_6$

$$W = \frac{\mathbb{1}}{2} - |G\rangle\langle G|$$

No. 10

$$W = \frac{\mathbb{1}}{2} - |G\rangle\langle G| - \frac{1}{2}\gamma_1^-\gamma_4^- - \frac{1}{2}\gamma_1^+\gamma_2^-\gamma_4^- - \frac{1}{2}\gamma_1^+\gamma_2^+\gamma_3^-\gamma_4^-$$

No. 11, H$_6$

$$W = \frac{\mathbb{1}}{2} - |G\rangle\langle G| - \frac{1}{2}\gamma_1^-\gamma_4^- - \frac{1}{2}\gamma_1^+\gamma_2^-\gamma_4^- - \frac{1}{2}\gamma_2^-\gamma_3^-\gamma_4^+ - \frac{1}{2}\gamma_1^-\gamma_2^+\gamma_3^-\gamma_4^+$$

No. 12, Y$_6$

$$W = \frac{\mathbb{1}}{2} - |G\rangle\langle G| - \frac{1}{2}\gamma_1^-\gamma_5^- - \frac{1}{2}\gamma_1^-\gamma_4^-\gamma_5^+ - \frac{1}{2}\gamma_1^+\gamma_4^-\gamma_6^- - \frac{1}{2}\gamma_1^+\gamma_4^+\gamma_5^-\gamma_6^-$$

No. 13, E$_6$

$$W = \frac{\mathbb{1}}{2} - |G\rangle\langle G| - \frac{1}{2}\gamma_1^-\gamma_5^- - \frac{1}{2}\gamma_1^-\gamma_5^+\gamma_6^- - \frac{1}{2}\gamma_1^+\gamma_5^-\gamma_6^-$$
$$- \frac{1}{4}\gamma_1^+\gamma_2^-\gamma_5^-\gamma_6^+ - \frac{1}{4}\gamma_1^-\gamma_4^-\gamma_5^-\gamma_6^+$$

No. 14, Cl$_6$

$$W = \frac{\mathbb{1}}{2} - |G\rangle\langle G| - \frac{1}{2}\gamma_1^-\gamma_4^- - \frac{1}{2}\gamma_1^+\gamma_3^-\gamma_6^- - \frac{1}{2}\gamma_1^-\gamma_4^+\gamma_6^-$$
$$- \frac{1}{4}\gamma_1^+\gamma_2^-\gamma_3^+\gamma_6^- - \frac{1}{4}\gamma_1^-\gamma_4^+\gamma_5^-\gamma_6^+ - \frac{1}{4}|011110\rangle\langle\cdot|$$

5.7 Witnesses

No. 15

$$W = \frac{1}{2}\mathbb{1} - |G\rangle\langle G| - \frac{1}{2}\gamma_1^-\gamma_2^-\gamma_3^+\gamma_5^- - \frac{1}{2}\gamma_1^-\gamma_2^-\gamma_3^-\gamma_5^+$$
$$-\frac{1}{3}\Big[|00\,\mathcal{T}(0011)\rangle\langle\cdot| + |01\,\mathcal{T}(0011)\rangle\langle\cdot| + |10\,\mathcal{T}(0011)\rangle\langle\cdot| + |010001\rangle\langle\cdot| + |010010\rangle\langle\cdot|$$
$$+ |010101\rangle\langle\cdot| + |010111\rangle\langle\cdot| + |011000\rangle\langle\cdot| + |011011\rangle\langle\cdot| + |011101\rangle\langle\cdot| + |011111\rangle\langle\cdot|$$
$$+ |100010\rangle\langle\cdot| + |100100\rangle\langle\cdot| + |100101\rangle\langle\cdot| + |100111\rangle\langle\cdot| + |101000\rangle\langle\cdot| + |101101\rangle\langle\cdot|$$
$$+ |101110\rangle\langle\cdot| + |101111\rangle\langle\cdot| + |110000\rangle\langle\cdot| + |110001\rangle\langle\cdot| + |110100\rangle\langle\cdot| + |110101\rangle\langle\cdot|$$
$$+ |111010\rangle\langle\cdot| + |111011\rangle\langle\cdot| + |111110\rangle\langle\cdot| + |111111\rangle\langle\cdot|\Big]$$

No. 16

$$W = \frac{1}{2}\mathbb{1} - |G\rangle\langle G| - \frac{1}{2}\gamma_1^-\gamma_5^- - \frac{1}{2}\gamma_1^-\gamma_5^+\gamma_6^- - \frac{1}{2}\gamma_1^+\gamma_5^-\gamma_6^-$$
$$-\frac{1}{4}\gamma_1^+\gamma_2^+\gamma_4^-\gamma_5^+\gamma_6^- - \frac{1}{4}\gamma_1^+\gamma_2^-\gamma_4^+\gamma_5^+\gamma_6^- - \frac{1}{4}\gamma_1^-\gamma_2^+\gamma_3^-\gamma_5^-\gamma_6^+ - \frac{1}{4}\gamma_1^+\gamma_2^-\gamma_3^-\gamma_5^-\gamma_6^+$$
$$-\frac{1}{4}\gamma_1^-\gamma_3^+\gamma_4^-\gamma_5^+\gamma_6^+ - \frac{1}{4}\gamma_1^-\gamma_3^-\gamma_4^+\gamma_5^+\gamma_6^+$$

No. 17

$$W = \frac{1}{2}\mathbb{1} - |G\rangle\langle G| - \frac{1}{2}\left(\gamma_2^+\gamma_5^- + \gamma_2^-\gamma_5^+\right)\gamma_6^-$$
$$-\frac{1}{2}\Big[|001101\rangle\langle\cdot| + |010110\rangle\langle\cdot| + |011010\rangle\langle\cdot| + |011110\rangle\langle\cdot| + |011111\rangle\langle\cdot|$$
$$+ |101101\rangle\langle\cdot| + |110110\rangle\langle\cdot| + |111010\rangle\langle\cdot| + |111110\rangle\langle\cdot| + |111111\rangle\langle\cdot|\Big]$$
$$- a\left(\gamma_2^+\gamma_5^+ + \gamma_2^-\gamma_5^-\right)\left(\gamma_3^+\gamma_4^- + \gamma_3^-\gamma_4^+\right)\gamma_6^-$$
$$- b\Big[|000110\rangle\langle\cdot| + |011000\rangle\langle\cdot| + |100010\rangle\langle\cdot| + |101010\rangle\langle\cdot| + |101110\rangle\langle\cdot|$$
$$+ |110000\rangle\langle\cdot| + |110100\rangle\langle\cdot| + |111100\rangle\langle\cdot|\Big],$$
$$a \approx 0.336,\ b \approx 0.163$$

No. 18, R_6

$$W = \frac{1}{2}\mathbb{1} - |G\rangle\langle G|$$
$$- \frac{1}{3}\Big[|\mathcal{T}(000011)\rangle\langle\cdot| + |\mathcal{T}(001011)\rangle\langle\cdot| + |\mathcal{T}(001101)\rangle\langle\cdot| + |\mathcal{T}(001111)\rangle\langle\cdot|\Big]$$
$$- a|111111\rangle\langle\cdot| - b\Big[|\mathcal{T}(011111)\rangle\langle\cdot| + |\mathcal{T}(010111)\rangle\langle\cdot|\Big]$$
$$- c\Big[|\mathcal{T}(001001)\rangle\langle\cdot| + |\mathcal{T}(011011)\rangle\langle\cdot| + |\mathcal{T}(010101)\rangle\langle\cdot|\Big],$$
$$a \approx 0.455, \ b \approx 0.363, \ c \approx 0.272$$

Note that expressions like $\mathcal{T}(001001)$ only sum over distinct translations, i.e. $|\mathcal{T}(011011)\rangle\langle\cdot| = |011011\rangle\langle\cdot| + |101101\rangle\langle\cdot| + |110110\rangle\langle\cdot|$.

No. 19

$$W = \frac{1}{2}\mathbb{1} - |G\rangle\langle G|$$
$$- \frac{1}{3}\Big[+ |\mathcal{T}(111110)\rangle\langle\cdot| + |\mathcal{T}(000011)\rangle\langle\cdot| + |\mathcal{T}(000101)\rangle\langle\cdot| + |\mathcal{T}(000111)\rangle\langle\cdot|$$
$$+ |\mathcal{T}(001001)\rangle\langle\cdot| + |\mathcal{T}(011011)\rangle\langle\cdot|$$
$$+ |001101\rangle\langle\cdot| + |010011\rangle\langle\cdot| + |010110\rangle\langle\cdot| + |011010\rangle\langle\cdot|$$
$$+ |011101\rangle\langle\cdot| + |011110\rangle\langle\cdot| + |100101\rangle\langle\cdot| + |101001\rangle\langle\cdot|$$
$$+ |101011\rangle\langle\cdot| + |101100\rangle\langle\cdot| + |101110\rangle\langle\cdot| + |110010\rangle\langle\cdot|$$
$$+ |110011\rangle\langle\cdot| + |110101\rangle\langle\cdot| + |111111\rangle\langle\cdot|\Big]$$

Again, expressions like $\mathcal{T}(001001)$ only sum over distinct translations.

6 Necessary and sufficient criteria for graph state entanglement

In this chapter, we characterize genuine multipartite entanglement for certain families of graph-diagonal states by providing necessary and sufficient criteria for its presence in these states and, by doing so, presenting cases in which the criterion introduced in Sec. 4 is necessary and sufficient for entanglement. We prove the presence of entanglement by applying some of the methods introduced in Sec. 5 and show the biseparability by constructing explicit biseparable decompositions.[1]

The graph-diagonal states that we consider are illustrated in Figure 6.1. Interest in graph-diagonal states is physically motivated: they occur naturally upon the decoherence of pure graph states [112–114], and, more importantly, any state can be brought into graph-diagonal form by local operations (cf. Sec. 2.5.2 and Ref. [76]). As local operations do not affect entanglement properties, this means that if the corresponding graph-diagonal state is entangled the original state was entangled. Thus, entanglement criteria for graph-diagonal states produce entanglement criteria for general states. Note that, for GHZ states, which are particular instances of graph states, the characterization of genuine multiparticle entanglement has already been solved [16].

This chapter is organized as follows: In Sec. 6.1 we consider the four-qubit cluster state and states diagonal in the corresponding basis. We provide a necessary and sufficient criterion for genuine multipartite entanglement, and, for states without multipartite entanglement, we provide an explicit decomposition into biseparable states. In Sec. 6.2 we discuss separability conditions for all five qubit graph states mixed with white noise. Again, we provide necessary and sufficient criteria for these families and explicit decompositions when the states are separable.

Then, in Sec. 6.3, we relate our results to the approach of Sec. 4 to characterize multiparticle entanglement via PPT mixtures (cf. Definition 26). Our result for the four-qubit cluster state implies that this criterion is necessary and sufficient for the four-qubit case. The question arises whether this is true in general. We argue that this may not be the case. Nevertheless, in Sec. 6.4 we discuss in detail the five-qubit Y-shaped graph state for which we can prove that the method of PPT mixtures does deliver a complete solution. In Sec. 6.5 we discuss generalizations of the Y-shaped graph to an arbitrary number of qubits. Our conclusions and a discussion of possible extensions of our work is presented in Sec. 6.6.

Before starting, we refer the reader unfamiliar with the notion of graph states to Sec. 2.5, where the necessary definitions are introduced. Note that all binary vectors like $|0011\rangle$ in the following are

---

[1]Reprinted excerpts with permission from Ref. [111], http://pra.aps.org/abstract/PRA/v84/i5/e052319. Copyright (2011) by the American Physical Society.

6 Necessary and sufficient criteria for graph state entanglement

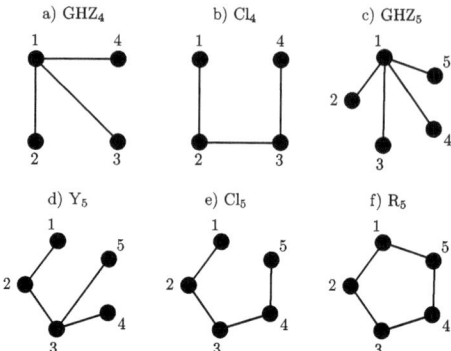

Figure 6.1: The graphs of the states discussed in this chapter. a) and c): star graphs corresponding to GHZ states. For GHZ-diagonal states with an arbitrary number of qubits, the problem of detecting multiparticle entanglement was already solved in Ref. [16]. b) The graph of the four-qubit cluster state. Theorem 50 gives a necessary and sufficient criterion for multipartite entanglement in states which are diagonal in the corresponding graph-state basis. d), e), f): five-qubit graph states $|Y_5\rangle, |Cl_5\rangle, |R_5\rangle$. In Sec. 6.2, we determine the border of separability when the graph diagonal generalizations of these states are mixed with white noise. For graph-diagonal states corresponding to the Y_5-graph, we also obtain a complete characterization of multiparticle entanglement (Theorem 57), which can be generalized to an arbitrary number of qubits (Theorem 59).

given in the graph state basis of the corresponding state under consideration. Also, for the sake of brevity, we write $|\psi\rangle\langle\cdot|$ instead of $|\psi\rangle\langle\psi|$.

6.1 Cluster-diagonal states of four qubits

In this section, we will derive a necessary and sufficient criterion for the presence of genuine multipartite entanglement in cluster-diagonal states of four qubits. Before proving our main result, we need two lemmas. The first one characterizes a set of entanglement witnesses for genuine multipartite entanglement, while the second one identifies a large class of biseparable quantum states that will simplify the search for biseparable decompositions.

6.1 Cluster-diagonal states of four qubits

Lemma 48. *The observables*

$$W_1 = \frac{1}{2} - |Cl_4\rangle\langle Cl_4| - \frac{1}{2}\frac{\mathbb{1}-g_1}{2}\frac{\mathbb{1}-g_4}{2}$$
$$= \frac{1}{2} - |0000\rangle\langle\cdot| - \frac{1}{2}\sum_{ij}|1ij1\rangle\langle\cdot|, \qquad (6.1)$$

$$W_2 = \frac{1}{2} - |0000\rangle\langle\cdot| - |1\alpha\beta 1\rangle\langle\cdot|, \qquad (6.2)$$

are entanglement witnesses for genuine multipartite entanglement. That is, $\mathrm{Tr}(\varrho W_k) < 0$ *implies the presence of genuine multipartite entanglement in* ϱ. *This holds for arbitrary* $\alpha, \beta \in \{0,1\}$ *in* W_2.

Proof. It was proven in Lemma 34 that W_1 is a witness [cf. Eq. (5.3)]. The fact that W_2 is a witness can be demonstrated in a similar way. It suffices to show that $(W_2)^{T_M} \geq 0$ for all possible bipartitions M. The operators $(W_2)^{T_M}$ are diagonal in the graph state basis. Thus, it is enough to show that $\langle ijkl|(W_2)^{T_M}|ijkl\rangle \geq 0$ holds for all elements of the graph state basis. This, however, is a direct consequence of Lemma 42 and Lemma 43. □

Lemma 49. *The quantum states*

$$\sigma = \frac{1}{2}(|ijkl\rangle\langle\cdot| + |\alpha\beta\gamma\delta\rangle\langle\cdot|) \qquad (6.3)$$

are biseparable, unless $i \neq \alpha$ *and* $l \neq \delta$ *both hold at the same time.*

Proof. First, we note that if $i \neq \alpha$ and $l \neq \delta$ the state is definitely not biseparable, since it is detected by a witness of the type W_2 (and also by W_1). Now, we show explicitly that all other states are biseparable. We can assume without loss of generality that $|ijkl\rangle = |0000\rangle$, since any $|ijkl\rangle$ can be transformed into $|0000\rangle$ by local transformations according to Eq. (2.70) and we neglect the normalization of σ. The first example is presented in great detail so as to demonstrate our methodology.

a) Consider the state $\sigma = |0000\rangle\langle\cdot| + |1000\rangle\langle\cdot|$. There are two ways to see that σ is biseparable with respect to the $A|BCD$ partition and we will discuss both of them, in order to illustrate the different methods.

(a1) The first method starts with the fact that for two qubits any mixture of two Bell states with equal weight (e.g., $\eta = |\Phi^+\rangle\langle\Phi^+| + |\Phi^-\rangle\langle\Phi^-|$) is separable [115]. The graph corresponding to a Bell state is the connected two-qubit graph. The four-qubit state σ can be considered as a separable mixture of the two Bell states $|00\rangle\langle\cdot| + |10\rangle\langle\cdot|$ on the first two qubits AB, where the qubits CD have been subsequently added to B via some local interaction. Clearly, the state σ remains biseparable between A and the rest of the qubits.

(a2) The second method uses Eq. (2.69) to write

$$\begin{aligned}\sigma &\propto (\mathbb{1}+g_2)(\mathbb{1}+g_3)(\mathbb{1}+g_4)\\ &\propto (\mathbb{1}+ZXZ\mathbb{1})(\mathbb{1}+\mathbb{1}ZXZ)(\mathbb{1}+\mathbb{1}\mathbb{1}ZX)\\ &\propto \underbrace{(\mathbb{1}+Z\mathbb{1}\mathbb{1}\mathbb{1})}_{\propto|0\rangle\langle 0|}\underbrace{(\mathbb{1}+g_2^{\text{red}})(\mathbb{1}+g_3)(\mathbb{1}+g_4)}_{\text{stabilizer state on qubits 2,3,4}} +\\ &\quad + \underbrace{(\mathbb{1}-Z\mathbb{1}\mathbb{1}\mathbb{1})}_{\propto|1\rangle\langle 1|}\underbrace{(\mathbb{1}-g_2^{\text{red}})(\mathbb{1}+g_3)(\mathbb{1}+g_4)}_{\text{stabilizer state on qubits 2,3,4}},\end{aligned} \qquad (6.4)$$

where $g_2^{\text{red}} = \mathbb{1}XZ\mathbb{1}$ denotes the restriction of the stabilizer g_2 to the qubits 2,3,4. In this form the state is clearly biseparable, since it is written as a sum of two terms, which are both biseparable with respect to the $A|BCD$-partition. This rewriting is possible, since in the expansion $\sigma \propto (\mathbb{1}+g_2)(\mathbb{1}+g_3)(\mathbb{1}+g_4)$ only the identity and one of the Pauli matrices (here: Z) occur on the first qubit as seen in Eq. (6.4). This statement holds for any Pauli operator. With these two methods at hand we now prove that the other states σ are also biseparable.

b) We consider $\sigma = |0000\rangle\langle\cdot| + |0100\rangle\langle\cdot|$. Using (a1) this is clearly separable with respect to the $A|BCD$ partition, but it is also separable with respect to the $B|ACD$-partition, as can be seen using the idea of (a2).

c) The state $\sigma = |0000\rangle\langle\cdot| + |1100\rangle\langle\cdot|$ is biseparable with respect to the $A|BCD$-partition according to (a1).

d) The state $\sigma = |0000\rangle\langle\cdot| + |1010\rangle\langle\cdot| \propto (\mathbb{1}+g_1g_3)(\mathbb{1}+g_2)(\mathbb{1}+g_4)$ is biseparable with respect to the $B|ACD$ partition, as can be seen using (a2).

e) The state $\sigma = |0000\rangle\langle\cdot| + |0110\rangle\langle\cdot|$ can be shown to be biseparable using the method of (a1) with qubits B and C as the Bell pair. Consequently, it is separable with respect to the $AB|CD$-partition.

f) Finally, we consider $\sigma = |0000\rangle\langle\cdot| + |1110\rangle\langle\cdot| \propto (\mathbb{1}+g_1g_2+g_2g_3+g_1g_3)(\mathbb{1}+g_4)$. First, using the method of (a2) one can directly calculate that this state is separable with respect to the $B|ACD$-partition. However, one can also apply the method of (a1): On the first three qubits, one can consider the state $\sigma = |000\rangle\langle\cdot| + |111\rangle\langle\cdot|$. This corresponds to a mixture of two three-qubit GHZ states, and it is known that such mixtures are always biseparable [16]. For σ only one qubit is added similar to (a1), so σ has to be biseparable, too. Up to symmetries these are all the relevant cases. □

We can now formulate and prove our main result. We denote the fidelities of the graph basis states as $F_{0000} = \langle 0000|\varrho|0000\rangle$ etc. We can then state:

Theorem 50. *A cluster-diagonal four-qubit state ϱ is biseparable, if and only if for all indices $\alpha, \beta, \gamma, \delta$*

$$2F_{\alpha\beta\gamma\delta} \le \sum_{i,j} F_{\alpha ij\delta} + \sum_{i,j} F_{\bar{\alpha} ij\delta} + \sum_{i,j} F_{\alpha ij\bar{\delta}} \qquad (6.5)$$

6.1 Cluster-diagonal states of four qubits

holds and for all indices $\alpha, \beta, \gamma, \delta, \mu, \nu$ the inequalities

$$2F_{\alpha\beta\gamma\delta} + 2F_{\bar{\alpha}\mu\nu\bar{\delta}} \leq \sum_{i,j} F_{\alpha ij\delta} + \sum_{i,j} F_{\bar{\alpha}ij\delta} + \sum_{i,j} F_{\alpha ij\bar{\delta}} + \sum_{i,j} F_{\bar{\alpha}ij\bar{\delta}} \tag{6.6}$$

are satisfied. Note that $\bar{\alpha} = 0$ for $\alpha = 1$ (and vice versa).

Before proving this result, let us interpret the conditions in Eqs. (6.5) and (6.6). In light of Lemma 49, Eq. (6.5) compares the weight of the state $|\alpha\beta\gamma\delta\rangle\langle\cdot|$ with the sum of the weights of all other states, which can be used to build a biseparable pair with $|\alpha\beta\gamma\delta\rangle\langle\cdot|$. If the overall state is biseparable, the first weight has to be smaller than the other weights, otherwise a decomposition with the methods of Lemma 49 cannot be found. The condition Eq. (6.6) then compares the weights of two states, $|\alpha\beta\gamma\delta\rangle\langle\cdot|$ and $|\bar{\alpha}\bar{\beta}\gamma\bar{\delta}\rangle\langle\cdot|$ (which, according to Lemma 49, do not constitute a separable pair) with all other weights. Using the normalization of the state, Eq. (6.6) can be rephrased as $F_{\alpha\beta\gamma\delta} + F_{\bar{\alpha}\mu\nu\bar{\delta}} \leq 1/2$, which has a natural meaning: if the weight of one "inseparable" pair exceeds all other weights, then the state cannot be separable.

Proof. We will use a shorthand notation for the sums such that Eq. (6.5), $2F_{0000} \leq \sum_{ij} F_{0ij0} + \sum_{ij} F_{1ij0} + \sum_{ij} F_{0ij1}$, is abbreviated as $2F_{0000} \leq \sum_{00} + \sum_{01} + \sum_{10}$.

We first have to show that if one of the conditions in Eqs. (6.5), (6.6) is violated, then ϱ is genuinely multipartite entangled. This follows directly from Lemma 48 since the conditions (6.5), (6.6) are nothing but a rewriting of $\text{Tr}(W_k\varrho) \geq 0$.

It remains to show that a state is biseparable, if Eqs. (6.5), (6.6) hold. Clearly, this is the difficult part. Our proof is split into four cases:

Case 1 — Let us first assume that the state ϱ acts only on the four-dimensional space spanned by the vectors $|1ij1\rangle$ and that the relevant four fidelities fulfill $F_{1\alpha\beta1} \leq F_{1\bar{\alpha}\beta1} + F_{1\alpha\bar{\beta}1} + F_{1\bar{\alpha}\bar{\beta}1}$ for all α, β. Then, the state ϱ is separable with respect to the $AB|CD$ partition. The reason is the following: a mixture of two-qubit Bell states, $\sigma = \lambda_{00}|\Phi^+\rangle\langle\Phi^+| + \lambda_{01}|\Phi^-\rangle\langle\Phi^-| + \lambda_{10}|\Psi^+\rangle\langle\Psi^+| + \lambda_{11}|\Psi^-\rangle\langle\Psi^-|$, is easily seen to be separable iff $\lambda_{\alpha\beta} \leq \lambda_{\bar{\alpha}\beta} + \lambda_{\alpha\bar{\beta}} + \lambda_{\bar{\alpha}\bar{\beta}}$ for all α, β [115]. The four-qubit state ϱ is nothing but a mixture of such Bell states between B and C, with qubits A and D added [see also case (a1) in the proof of Lemma 49].

Case 2 — Now we assume that equality holds for one of the conditions of Eq. (6.5). Without loss of generality, we assume that $2F_{0000} = \sum_{00} + \sum_{01} + \sum_{10}$ while the other conditions in Eqs. (6.5), (6.6) are fulfilled, but not necessarily with equality.

In this case, Eq. (6.6) becomes $F_{1\alpha\beta1} \leq F_{1\bar{\alpha}\beta1} + F_{1\alpha\bar{\beta}1} + F_{1\bar{\alpha}\bar{\beta}1}$, the same relation discussed in Case 1. If we consider now the projection ϱ^R of the original state ϱ on the four-dimensional space spanned by the vectors $|1ij1\rangle$, then it is clear that this state ϱ^R is separable according to Case 1. It remains to show that the orthogonal part $\varrho - \varrho^R$ is separable too. For this part, we have $F_{0000} = F_{0110} + F_{0010} + F_{0100} + \sum_{01} + \sum_{10}$, so it can be directly decomposed with the help of Lemma 49, by using all possible combinations of the type $|0000\rangle\langle\cdot| + |\alpha\beta\gamma\delta\rangle\langle\cdot|$. This finishes the proof of Case 2.

Case 3 — In this case, we assume that equality holds for one of the conditions of Eq. (6.6): $2F_{0000} + 2F_{1111} = \sum_{00} + \sum_{01} + \sum_{10} + \sum_{11}$. The other inequalities in Eqs. (6.5), (6.6) are satisfied, but not

111

necessarily with equality. We rewrite the equality as $(2F_{0000} - \sum_{00}) + (2F_{1111} - \sum_{11}) = \sum_{01} + \sum_{10}$. Using this together with the inequalities given by Eq. (6.5) we can also deduce the conditions $2F_{1111} \geq \sum_{11}$ and $2F_{0000} \geq \sum_{00}$.

Now, we can decompose ϱ as follows: Consider the space spanned by $|1ij1\rangle$, and a state σ^- with a fidelity F'_{1111} that equals $F'_{1111} = \sum_{11} - F_{1111} = F_{1101} + F_{1011} + F_{1001}$ and the other fidelities obey $F'_{1ij1} = F_{1ij1}$. This state is separable according to Case 1, since $F'_{1111} = F'_{1101} + F'_{1011} + F'_{1001}$. The restriction ϱ^R of ϱ onto the four-dimensional subspace is now given by $\varrho^R = \sigma^- + (F_{1111} - F'_{1111})|1111\rangle\langle\cdot| = \sigma^- + (2F_{1111} - \sum_{11})|1111\rangle\langle\cdot|$. We can make a similar construction on the space spanned by $|0ij0\rangle$ with a separable state σ^+. A projector onto $|0000\rangle$ with weight $(2F_{0000} - \sum_{00})$ will remain.

Therefore, we can decompose ϱ into the two separable states σ^- and σ^+ on the four-dimensional spaces and a remaining state η. The state η has only two contributions on the two four-dimensional spaces, which have the fidelities $F^\eta_{0000} = 2F_{0000} - \sum_{00}$ and $F^\eta_{1111} = 2F_{1111} - \sum_{11}$. From our assumption, it follows that η fulfills $F^\eta_{0000} + F^\eta_{1111} = \sum_{01} + \sum_{10}$. This remaining state η can then be decomposed using states of the form $\sigma = |1111\rangle\langle\cdot| + |1kl0\rangle\langle\cdot|$ and $\sigma = |1111\rangle\langle\cdot| + |0kl1\rangle\langle\cdot|$ etc.

Case 4 — Let us finally discuss the case where equality holds for none of the conditions of Eqs. (6.5) and (6.6). We consider the state

$$\varrho^{\text{new}} = \varrho - \varepsilon\sigma, \qquad (6.7)$$

where σ is one of the separable states from Lemma 49. Since $\varrho = \varrho^{\text{new}} + \varepsilon\sigma$, the state ϱ is separable, if ϱ^{new} is separable and positive.

The idea is to choose possible biseparable states σ and subtract them step by step such that ϱ^{new} remains positive. Note that during these subtractions, the inequalities (6.5), (6.6) become tighter. But one does not have to worry that they become violated: If they become violated, at some point equality must have held in one of the two Eqs. (6.5) and (6.6) first, while the other conditions still hold. This means that at this point ϱ^{new} (and hence ϱ) is separable, according to Cases 2 and 3.

What can be achieved with the iterative subtractions? First, by subtracting the biseparable states $\sigma = |0ij0\rangle\langle\cdot| + |0kl0\rangle\langle\cdot|$ one can set three of the four fidelities $F_{0\alpha\beta 0}$ to zero. Similarly, in each of the sets $\{F_{0\alpha\beta 1}\}$, $\{F_{1\alpha\beta 0}\}$ and $\{F_{1\alpha\beta 1}\}$ three fidelities can be made to vanish, such that overall only four F_{ijkl} are nonzero. The structure of the fidelities is now such that all the sums in Eqs. (6.5), (6.6) contain only a single term. Then, however, Eq. (6.6) must either be violated for some set of indices or equality must hold. □

Using this theorem, one finds that cluster states mixed with white noise, $\varrho(p) = p|Cl_4\rangle\langle Cl_4| + (1-p)\mathbb{1}/16$, are entangled if and only if $p > 5/13$. This confirms a numerically established threshold from Table 4.1 in Sec. 4.4.

Furthermore, the theorem demonstrates that for cluster-diagonal states there are effectively only two entanglement witnesses, namely the ones from Lemma 48. It is interesting to compare this with the results of Ref. [16], where a necessary and sufficient criterion for GHZ-diagonal states was found. This criterion can be interpreted in the sense that for GHZ-diagonal states (of an arbitrary number of qubits) only one entanglement witness is relevant, namely $W = \mathbb{1}/2 - |GHZ_N\rangle\langle GHZ_N|$. For cluster states, the witness $W = \mathbb{1}/2 - |Cl_4\rangle\langle Cl_4|$ is not optimal, since both of the witnesses in Lemma 48

6.2 Five-qubit graph states

are finer. One can expect that for more complicated graph states of more qubits, a significantly higher number of witnesses is relevant, hence a complete classification becomes difficult.

6.2 Five-qubit graph states

In this section, we derive optimal criteria for all five-qubit graph states mixed with white noise. Doing this demonstrates that the witnesses obtained with the PPT approach of Sections 4 and 5 have the highest white-noise tolerance possible and are in this sense optimal. In the next section, however, we will argue that the success of the PPT approach in finding optimal witnesses might be specific to these states. Nevertheless, we present a full solution of the cluster state Y_5 in Theorem 57 of Sec. 6.4.

6.2.1 The state Y_5

For the five-qubit Y_5 state (cf. Fig. 6.1) mixed with white noise we find:

Lemma 51. *The state*

$$\varrho(p) = p|Y_5\rangle\langle Y_5| + (1-p)\frac{\mathbb{1}}{32} \tag{6.8}$$

is genuinely multipartite entangled if and only if $p > 9/25 = 0.36$.

Proof. First, for the case that $p > 9/25$ the state $\varrho(p)$ is detected by the witness of Eq. (5.11) constructed according to Lemma 36,

$$W_{Y_5} = \frac{\mathbb{1}}{2} - |Y_5\rangle\langle Y_5| - \frac{1}{16}[(\mathbb{1} - g_1)(\mathbb{1} - g_4)(\mathbb{1} + g_5) \\ + (\mathbb{1} - g_1)(\mathbb{1} + g_4)(\mathbb{1} - g_5) + (\mathbb{1} - g_1)(\mathbb{1} - g_4)(\mathbb{1} - g_5)] \tag{6.9}$$

and, hence, genuine multipartite entangled.

In the other direction, we first have to identify the separable states as we did in Lemma 49. In fact, for many states this lemma can directly be generalized. For instance, the state $\sigma = |00000\rangle\langle\cdot| + |ijk00\rangle\langle\cdot|$ is biseparable, since for the four-qubit cluster state $\sigma' = |0000\rangle\langle\cdot| + |ijk0\rangle\langle\cdot|$ is separable, and the fifth qubit is added as in case (a1) in the proof of Lemma 49 . In fact, the only combinations which are *not* separable are of the form $\chi_1 = |00000\rangle\langle\cdot| + |1jk10\rangle\langle\cdot|$ and $\chi_2 = |00000\rangle\langle\cdot| + |1jkl1\rangle\langle\cdot|$. Note that the state $\sigma = |00000\rangle\langle\cdot| + |0jk11\rangle\langle\cdot|$ is biseparable, because it can be considered as a separable four-qubit GHZ state on BCDE where one qubit is added [see case f) in the proof of Lemma 49].

The state at the critical value of p is

$$\varrho \propto 19|00000\rangle\langle\cdot| + \sum_{ijklm \neq 00000} |ijklm\rangle\langle\cdot|, \tag{6.10}$$

and it remains to show that this state is separable. First, the state

$$\varrho' = 19|00000\rangle\langle\cdot| + \Big(\sum_{ijklm\neq 00000}|ijklm\rangle\langle\cdot| - \sum_{ij}|1ij10\rangle\langle\cdot| - \sum_{ij}|1ij01\rangle\langle\cdot| - \sum_{ij}|1ij11\rangle\langle\cdot|\Big) \quad (6.11)$$

is biseparable, since in the sums in the brackets exactly 19 terms remain, and a decomposition with σ from above is then straightforward. The remaining term $\varrho - \varrho' = \sum_{ij}|1ij10\rangle\langle\cdot| + \sum_{ij}|1ij01\rangle\langle\cdot| + \sum_{ij}|1ij11\rangle\langle\cdot|$ is also clearly separable, since the sum of any two of the occurring states is separable. $\square$

6.2.2 The linear cluster state Cl_5

For the five-qubit linear cluster state $|Cl_5\rangle$ [cf. Fig. 6.1 e)] mixed with white noise, the threshold is the same as for the Y_5-state:

Lemma 52. *The state*

$$\varrho(p) = p|Cl_5\rangle\langle Cl_5| + (1-p)\frac{1}{32} \quad (6.12)$$

is genuinely multipartite entangled if and only if $p > 9/25 = 0.36$.

Proof. First, the witness constructed according to Lemma 36,

$$W_{Cl5} = \frac{1}{2} - |Cl_5\rangle\langle Cl_5| - \frac{1}{32}\big[4(\mathbb{1}-g_1)(\mathbb{1}-g_5) + (\mathbb{1}+g_1)(\mathbb{1}-g_2)(\mathbb{1}-g_5) + (\mathbb{1}-g_1)(\mathbb{1}-g_4)(\mathbb{1}+g_5)\big], \quad (6.13)$$

detects the state for $p > 9/25$, proving one part of the claim.

For the other direction, we have to again identify the biseparable states. First, in a generalization of Lemma 49, states of the form $\sigma = |00000\rangle\langle\cdot| + |ijklm\rangle\langle\cdot|$ are separable, unless they are of the form $\chi_1 = |00000\rangle\langle\cdot| + |1jk10\rangle\langle\cdot|$, $\chi_2 = |00000\rangle\langle\cdot| + |01jk1\rangle\langle\cdot|$, or $\chi_3 = |00000\rangle\langle\cdot| + |1jkl1\rangle\langle\cdot|$. There are 16 terms of this type which are not biseparable.

The state at $p = 9/25$ is given by $\varrho = 19|00000\rangle\langle\cdot| + \sum_{ijklm\neq 00000}|ijklm\rangle\langle\cdot|$ (up to a normalization). Generalizing Lemma 49 we can subtract many pairs of terms such that what remains is to show that

$$\varrho' = 4|00000\rangle\langle\cdot| + \sum_{ij}|1ij10\rangle\langle\cdot| + \sum_{ij}|01ij1\rangle\langle\cdot| + \sum_{ijk}|1ijk1\rangle\langle\cdot| \quad (6.14)$$

6.2 Five-qubit graph states

is separable. In order to do so, we consider the four states

$$\eta_1 = |00000\rangle\langle\cdot| + |01001\rangle\langle\cdot| + |10010\rangle\langle\cdot| + |11011\rangle\langle\cdot|,$$
$$\eta_2 = |00000\rangle\langle\cdot| + |01011\rangle\langle\cdot| + |11010\rangle\langle\cdot| + |10001\rangle\langle\cdot|,$$
$$\eta_3 = |00000\rangle\langle\cdot| + |01101\rangle\langle\cdot| + |11110\rangle\langle\cdot| + |10011\rangle\langle\cdot|,$$
$$\eta_4 = |00000\rangle\langle\cdot| + |01111\rangle\langle\cdot| + |10110\rangle\langle\cdot| + |11001\rangle\langle\cdot|. \quad (6.15)$$

The state η_1 is separable for the following reason: it is known that the four-qubit Smolin state $\sigma = |\Phi^+\rangle\langle\Phi^+|_{AB} \otimes |\Phi^+\rangle\langle\Phi^+|_{A'B'} + |\Phi^-\rangle\langle\Phi^-|_{AB} \otimes |\Phi^-\rangle\langle\Phi^-|_{A'B'} + |\Psi^+\rangle\langle\Psi^+|_{AB} \otimes |\Psi^+\rangle\langle\Psi^+|_{A'B'} + |\Psi^-\rangle\langle\Psi^-|_{AB} \otimes |\Psi^-\rangle\langle\Psi^-|_{A'B'}$ is separable with respect to the $AA'|BB'$ partition [116]. The state η_1 is simply a Smolin state between the qubits $ABDE$, where the qubit C has been added [see case (a1) in Lemma 49]. Therefore, it is separable with respect to the $AE|BCD$-partition. Similarly, η_2 is a Smolin state up to local unitary operations and therefore separable with respect to the same partition.

It can be directly verified that the state η_3 is PPT with respect to the $BD|ACE$ partition. This implies separability via the following argument: For the considered partition, η_3 is acting on a 4×8 (effectively 4×4) space. The PPT entangled states in this scenario have at least rank of five [117]. Hence, η_3, which is of rank four, must be separable with respect to the partition[2]. Similarly, η_4 is separable with respect to the $BD|ACE$-partition.

So we can write

$$\varrho' = \sum_{k=1}^{4} \eta_k + \sum_{ij} |1i1j1\rangle\langle\cdot| \quad (6.16)$$

where the sum of the remaining four projectors is clearly separable according to Lemma 49. This finishes the proof. □

6.2.3 The ring cluster state R_5

For the five-qubit ring cluster state mixed with white noise the separability problem can be solved as follows:

Lemma 53. *The state*

$$\varrho(p) = p|R_5\rangle\langle R_5| + (1-p)\frac{1}{32} \quad (6.17)$$

is genuinely multipartite entangled if and only if $p > 7/19 \approx 0.368.$

---

[2] Alternatively, one can see the separability of η_3 as follows: Applying a local complementation on qubit 2 and then on qubit 1 exchanges qubits 1 and 2. Similarly, a local complementation first on qubit 4 and then on qubit 5 exchanges qubits 4 and 5. The signs of the states $|ijklm\rangle$ in the graph-state basis are not invariant under these transformations. Applying the rules of a local complementation (cf. Sec. 2.5.3 and Ref. [64]), a complementation on qubit a flips the signs in the neighborhood $\mathcal{N}(a)$ if and only if the sign on a is -1. With this rule, one sees that after a complementation on the qubits 2, then 1, then 4, then 5 the state is like a Smolin state between the qubits ABDE, and the qubit C is connected to the qubits A and E, so it is separable with respect to the $BD|ACE$ partition. The same argument can be applied to η_4.

6 Necessary and sufficient criteria for graph state entanglement

Proof. Due to the symmetry of this state, it is convenient for our discussion to define $\mathcal{T}(x)$ as the sum over all five translations of the term x, corresponding to a rotation of the ring graph. For example, $\mathcal{T}(|00001\rangle\langle\cdot|) = |00001\rangle\langle\cdot| + |00010\rangle\langle\cdot| + |00100\rangle\langle\cdot| + |01000\rangle\langle\cdot| + |10000\rangle\langle\cdot|$.

A witness for the ring cluster state of five qubits can be found in Sec. 5.7 (state No. 8). It is given by

$$W_{R5} = 3\big[\mathcal{T}(|00001\rangle\langle\cdot|) + \mathcal{T}(|00101\rangle\langle\cdot|) + \mathcal{T}(|00111\rangle\langle\cdot|)\big]$$
$$+ \big[\mathcal{T}(|00011\rangle\langle\cdot|) + \mathcal{T}(|01011\rangle\langle\cdot|) + \mathcal{T}(|01111\rangle\langle\cdot|)\big]$$
$$- |11111\rangle\langle\cdot| - 3|00000\rangle\langle\cdot|. \tag{6.18}$$

This witness detects the entanglement in the state for $p > 7/19$, proving one direction of the claim.

For the other direction, we have to identify separable states. First, states like $\sigma_1 = |00000\rangle\langle\cdot| + |00001\rangle\langle\cdot|$, $\sigma_2 = |00000\rangle\langle\cdot| + |00101\rangle\langle\cdot|$ and $\sigma_3 = |00000\rangle\langle\cdot| + |00111\rangle\langle\cdot|$ are clearly separable in analogy to Lemma 49: σ_1 is separable in analogy to case (a2), σ_2 and σ_3 can be considered as states on the qubits $BCDE$ which are separable with respect to the $D|BCE$ partition [cases d) and f) in Lemma 49], where the qubit A is added by a local transformation. Furthermore, the states

$$\eta_1 = |00000\rangle\langle\cdot| + |01100\rangle\langle\cdot| + |11010\rangle\langle\cdot| + |10110\rangle\langle\cdot|,$$
$$\eta_2 = |00000\rangle\langle\cdot| + |11000\rangle\langle\cdot| + |00110\rangle\langle\cdot| + |11110\rangle\langle\cdot|,$$
$$\eta_3 = |00000\rangle\langle\cdot| + |11000\rangle\langle\cdot| + |10111\rangle\langle\cdot| + |01111\rangle\langle\cdot|,$$
$$\eta_4 = |00000\rangle\langle\cdot| + |11010\rangle\langle\cdot| + |01101\rangle\langle\cdot| + |10111\rangle\langle\cdot| \tag{6.19}$$

are also separable. The state η_1 is separable with respect to the $BC|ADE$-partition, as can be seen from the separability properties of the Smolin state (similar to the state η_1 defined for the linear cluster state Cl_5 above). η_2 is PPT with respect to the $BC|ADE$-partition, and hence separable (due to a rank argument as in the proof of Lemma 52.). The separability of η_3 (and η_4) can be inferred from the fact that they are PPT with respect to the $CE|ABD$ (and $AC|BDE$) partition.

The state at $p = 7/19$ is given by $\varrho = 59|00000\rangle\langle\cdot| + 3\sum_{ijklm \neq 00000} |ijklm\rangle\langle\cdot|$ which can be written as

$$\varrho = 3\sum_{k=1}^{3} \mathcal{T}(\sigma_k) + \frac{14}{20}\sum_{k=1}^{4} \mathcal{T}(\eta_k) + \varrho' \tag{6.20}$$

with

$$\varrho' = \frac{1}{5}\mathcal{T}(|00011\rangle\langle\cdot|) + \frac{1}{5}\mathcal{T}(|01011\rangle\langle\cdot|) + \frac{1}{5}\mathcal{T}(|01111\rangle\langle\cdot|) + 3|11111\rangle\langle\cdot|. \tag{6.21}$$

This state, however, can directly be decomposed in terms of the σ_k with all signs inverted. □

6.3 Connection with the theory of PPT mixtures

In order to place our results within a wider framework, we discuss possible connections with the theory of PPT mixtures, introduced in Definition 26 of Sec. 4. With respect to the results reported here, it is remarkable that all the witnesses used in this chapter [Lemma 48 and Eqs. (6.9), (6.13) and (6.18)] were derived from the theory of PPT mixtures. Any PPT mixture within the considered subclass (e.g., the cluster-diagonal states) will therefore fulfill the conditions set by the witnesses [e.g., Eqs. (6.5), (6.6)] and must be biseparable. In other words, we have shown that for the families of graph-diagonal states considered here, biseparability is equivalent to being a PPT mixture.

This leads to the question, whether it is generally true that graph-diagonal states are biseparable if and only if they are PPT mixtures. If this conjecture were true, it would solve the problem of characterizing multiparticle entanglement for a huge class of states with an arbitrary number of qubits. Moreover, for graph-diagonal states the problem can be solved with linear programming, which is significantly simpler than semidefinite programming (cf. Lemma 33) and which could deal with larger qubit systems. However, there is evidence that the conjecture is not correct, as we explain in the following.

First, note that when looking for a decomposition of a graph-diagonal state into biseparable (or PPT) states, one can assume that the terms in the decomposition are also graph-diagonal. If one finds a decomposition where this is not the case, one can always apply the local operations described in Sec. 2.5.2 and Ref. [76], which turn a general state into a graph-diagonal state without changing its fidelities with respect to the graph basis vectors. The graph-diagonal state is invariant under these operations, but terms in the decomposition which are not diagonal become diagonal after application of these operations. Since this operation is local, the state remains biseparable or PPT.

Therefore, if any graph-diagonal state which is PPT with respect to a given bipartition, is also separable with respect to the same bipartition, the conjecture would be correct. However, this is not always the case. Examples can be given from bound entangled states known in the literature [118, 119]. For instance, consider the four-qubit cluster-diagonal state

$$\tilde{\varrho} = \frac{1}{6}(|0011\rangle\langle\cdot| + |1001\rangle\langle\cdot| + |1110\rangle\langle\cdot| + |0101\rangle\langle\cdot| + |0111\rangle\langle\cdot| + |0110\rangle\langle\cdot|). \quad (6.22)$$

Although this state is PPT with respect to the $AD|BC$-bipartition, it is entangled with respect to the same bipartition. This follows from Ref. [118], in which the four-qubit state $\hat{\varrho}$, a mixture of Bell states

$$\hat{\varrho} = \frac{1}{6}(|\Phi^+\rangle\langle\cdot|_{AB} \otimes |\Psi^-\rangle\langle\cdot|_{A'B'} + |\Psi^+\rangle\langle\cdot|_{AB} \otimes |\Psi^+\rangle\langle\cdot|_{A'B'} + |\Psi^-\rangle\langle\cdot|_{AB} \otimes |\Phi^-\rangle\langle\cdot|_{A'B'}$$
$$+ |\Phi^-\rangle\langle\cdot|_{AB} \otimes |\Psi^+\rangle\langle\cdot|_{A'B'} + |\Phi^-\rangle\langle\cdot|_{AB} \otimes |\Psi^-\rangle\langle\cdot|_{A'B'} + |\Phi^-\rangle\langle\cdot|_{AB} \otimes |\Phi^-\rangle\langle\cdot|_{A'B'},) \quad (6.23)$$

is shown to be PPT, but still entangled, with respect to the $AA'|BB'$-partition. Since the Bell states can be interpreted as two-qubit graph states, this is a graph-diagonal state. Adding a connection between the qubits B and B' via a controlled phase gate leads to the four-qubit cluster-diagonal state $\tilde{\varrho}$ in Eq. (6.22) which is PPT for the $AD|BC$-partition, but nevertheless entangled. Similar examples could be constructed for higher numbers of qubits [119]. This demonstrates that for higher numbers of qubits there might be graph-diagonal states which are PPT mixtures, but nevertheless genuinely multipartite entangled.

6.4 PPT mixtures and the five-qubit Y_5 state

In the previous section, we have argued that, in general, one cannot expect that the criterion of PPT mixtures is necessary and sufficient for entanglement in graph-diagonal states. In this section, however, we show that for graph-diagonal states associated to the five-qubit Y_5 graph, the criterion of PPT mixtures is, in fact, necessary and sufficient for entanglement.

The basic idea of our proof is that for the Y_5 state, bound entangled states such as those given in Eqs. (6.22), (6.23) play no role in the decomposition. To start, note that the state $\tilde{\varrho}$ in Eq. (6.22), despite being entangled for $AD|BC$, is biseparable and a decomposition can directly be written down with the help of Lemma 49. This highlights an interesting detail in the proof of Theorem 50. For the biseparable decompositions identified in Lemma 49, only the bipartitions $A|BCD$ (and permutations) and $AB|CD$ have been used, but not the bipartitions $AD|BC$ and $AC|BD$. Interestingly, there is a fundamental difference between these types of bipartitions. For the first set, the entanglement between the two partitions in the pure graph state $|Cl_4\rangle$ is equal to one Bell-pair (or one e-bit). This can be seen from the Schmidt decomposition of $|Cl_4\rangle$ with respect to that partition (where the Schmidt coefficients are both $1/\sqrt{2}$). Alternatively, this follows from the structure of the graph, since, after suitable transformations which are local for the given bipartition, there is only one connection between the parties. In the second set, namely the bipartitions $AD|BC$ and $AC|BD$, the entanglement between the partitions is equal to two Bell pairs. Consequently, we refer to the first type of bipartitions as 1BP and the second type as 2BP.

We can now formulate a fundamental observation linking PPT to separability. If we have a graph-diagonal state and a 1BP bipartition, then the PPT criterion is clearly necessary and sufficient for separability since, after suitable local operations, the state can be viewed as a two-qubit state[3]. On the other hand, for a 2BP bipartition, this is definitely not the case, as the examples in Eqs. (6.22), (6.23) demonstrate. We formulate this as follows:

Corollary 54. *For any biseparable cluster-diagonal state of four qubits there is a decomposition using 1BP bipartitions only. Consequently, when looking for a decomposition for a given four-qubit cluster-diagonal state, it suffices to consider 1BP bipartitions only.*

This statement directly follows from the proof of Theorem 50, since only 1BP bipartitions have been used there. It is straightforward to generalize this slightly as follows:

---

[3]To give a precise argument, consider a three-qubit graph-diagonal state ϱ using the linear graph 1—2—3 which is PPT with respect to the $A|BC$-bipartition. After a controlled phase gate between qubits 2,3, which is a local operation for the $A|BC$-bipartition, the state is transformed to $\tilde{\varrho} = \varrho_{AB}^+ \otimes |+\rangle\langle+|_C + \varrho_{AB}^- \otimes |-\rangle\langle-|_C$ where $|\pm\rangle = (|0\rangle \pm |1\rangle)/\sqrt{2}$ and the states $\varrho_{AB}^\pm$ are two-qubit graph-diagonal states for the graph 1—2. Since one can deterministically prepare ϱ_{AB}^+ and ϱ_{AB}^- by measuring X on the third qubit, both the $\varrho_{AB}^\pm$ must also be PPT and hence separable. This demonstrates that the original state ϱ was also separable. A similar argument is used in the proof of Lemma 56.

6.4 PPT mixtures and the five-qubit Y_5 state

Lemma 55. *Let ϱ be a four-qubit graph-diagonal state for an arbitrary graph, which is PPT with respect to a given bipartition. Then, ϱ can be written as a PPT mixture using 1BP bipartitions only.*

Proof. First, note that the statement is only non-trivial if the given bipartition is 2BP. Furthermore, note that up to local complementations (or local unitaries) there are only two different graphs, the GHZ_4-graph and the linear cluster graph Cl_4. For the GHZ_4-graph any bipartition is 1BP. For the cluster graph, however, being PPT for the given partition implies that the state is biseparable, since then the expectation values of the witnesses in Lemma 48 are non-negative. Then the claim follows from Corollary 54. □

In order to apply similar ideas to the Y_5 state, we need to generalize the above statement to five qubits. For five qubits, one can similarly consider 1BP and 2BP bipartitions. There are no bipartitions with three Bell pairs, as this would require at least six qubits.

Lemma 56. *Consider a connected five-qubit graph and a two- vs. three-qubit bipartition where one of the qubits in the three-qubit part of the bipartition is connected with only one other qubit in the same three-qubit part. Let ϱ be a graph-diagonal five-qubit state being PPT for the given partition. Then, ϱ is a PPT mixture using 1BP partitions only.*

First, to give an example where the condition on the graph holds, consider the Y_5 graph in Fig. 6.1 d) and the $ACE|BD$ (or $135|24$) partition. Then, the qubit E (or 5) is connected only with one qubit in the same part of the partition, namely the qubit C (or 3). Thus, the condition on the graph is fulfilled. Note that in this case, the bipartition is a 2BP bipartition, so the statement of the lemma is not trivial.

Proof. To prove Lemma 56, we assume without loss of generality that the bipartition is the $AB|CDE$ bipartition and E is the singular qubit connected only with qubit D. By a suitable local transformation (acting on DE only), one can decouple qubit E from the rest. This means that the state ϱ is transformed to $\hat{\varrho} = \varrho^+ \otimes |+\rangle\langle+|_E + \varrho^- \otimes |-\rangle\langle-|_E$ where the $\varrho^\pm$ are unnormalized states on the qubits $ABCD$. Note that $|+\rangle$ and $|-\rangle$ here are the eigenstates of the Pauli operator X. Since $\hat{\varrho}$ is PPT with respect to the $AB|CDE$ partition, the states $\varrho^\pm$ must also be PPT with respect to the $AB|CD$ partition. Otherwise, it would be possible to generate non-positive partial transpose (NPT) entanglement from a PPT state by measuring E and distinguishing between ϱ^+ and ϱ^-. This is known to be impossible [120]. Hence, according to Lemma 55, the states ϱ^+ and ϱ^- form PPT mixtures with respect to 1BP bipartitions on the qubits $ABCD$. Reconnecting the qubit E on the side of D in a 1BP bipartition on $ABCD$ leads to a bipartition on five qubits, which is 1BP, even if E is again connected with D. This immediately induces a PPT mixture of ϱ, where only 1BP bipartitions occur in the decomposition. □

We can now formulate our main result for the Y_5 state where all two- vs. three-qubit bipartitions (2-3-bipartitions) are either 1BP or fulfill the conditions of Lemma 56.

Theorem 57. *A Y_5-graph-diagonal state is biseparable, if and only if it is a PPT mixture.*

Proof. Clearly, a biseparable state is also a PPT mixture, which proves one direction of the claim. Concerning the other direction, let us consider a PPT mixture and recall that if a PPT mixture is graph-diagonal, then the terms in the mixture can be chosen to be graph-diagonal as well (cf. Sec. 5.1). We will argue that the terms belonging to the 2BP bipartitions in the PPT mixture of the state Y_5 can be written as mixtures of 1BP bipartitions. For this we will make use of Lemma 56.

The only candidates for 2BP bipartitions are the 2-3-bipartitions, as the 1-4-bipartitions are automatically 1BP. For the Y_5-graph, several 2-3-bipartitions are in 2BP, however, all fulfill the conditions of Lemma 56: The $ACE|BD$-bipartition has already been discussed and the $AE|BCD$-bipartition satisfies the condition directly. The $BC|ADE$-bipartition is 2BP and does not fulfill the condition directly. Nevertheless, after a local complementation on qubit C and then on qubit E, the qubit D is left connected only with qubit E so that it meets the conditions of Lemma 56. The same sequence of local complementations can be applied to the $AC|BDE$-bipartition to show that it also meets the conditions of Lemma 56. These are, up to symmetries, all of the 2BP bipartitions. This implies that all 2BP bipartitions of Y_5 meet the conditions of Lemma 56 and thus the PPT criterion is necessary and sufficient to demonstrate multiparticle entanglement. This proves the claim. □

From the proof of Theorem 10 it also follows that the search for the decomposition into PPT states can be restricted to 1BP bipartitions. In practice, one can easily modify the existing algorithms to consider 1BP bipartitions only [101], which would even make the numerical program simpler.

An extension of this theorem to other five-qubit graphs is not straightforward. For instance, for the linear cluster graph [Fig. 6.1 e)] the bipartition $BD|ACE$ is 2BP but does not fulfill the conditions of Lemma 56 even after local complementation. However, this bipartition seems to be relevant in the decomposition, since it is used in η_3 of Eq. (6.15).

6.5 Generalizations to more than five particles

So far, we have investigated the separability problem for graph-diagonal states with up to five qubits and found solutions for many important cases. In this section we provide two examples that demonstrate how our results can be used to investigate entanglement in graph-diagonal states with even larger number of qubits.

6.5.1 A generalization to Y_N-states

In our first example, we consider the Y_N-state, a generalization of the Y_5-state, which we show in Fig. 6.2a). For this family of states, we can generalize Theorem 57 and show that the criterion of PPT mixtures is necessary and sufficient. We need the following lemma:

Lemma 58. *Let ϱ be a Y_N-graph-diagonal state with $N \geq 5$ and consider a 2BP bipartition. Then, if ϱ is PPT with respect to that partition, it can be written as a PPT mixture using 1BP partitions only.*

6.5 Generalizations to more than five particles

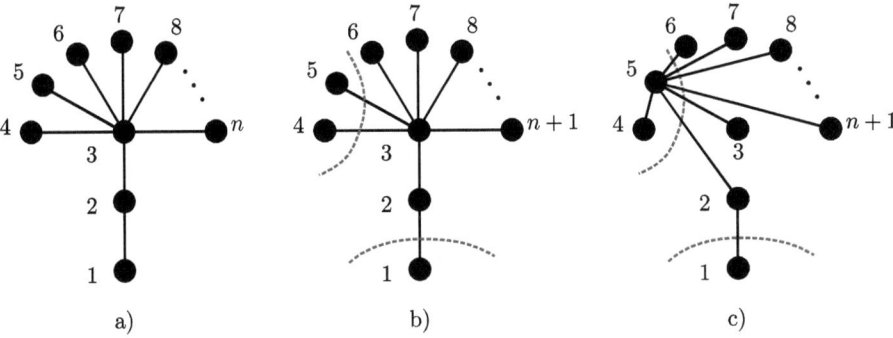

Figure 6.2: a) The graph of a Y_N-state, a possible generalization of the Y_5-state. b) A possible 2BP bipartition, here we have chosen $i = 5$ and $j = 4$. c) After local complementation on the qubits 3 and $i = 5$, the qubit 5 is the "central" qubit. The qubits 4 and 5 then fulfill the condition for fulfill Lemma 56. See text for further discussion.

Proof. We prove the statement by induction. The base case for the induction, $n = 5$, has already been proven. For the inductive step, consider the Y_{n+1}-graph and a 2BP partition [see Fig. 6.2 b)]. We denote the two parts of the bipartition as M and $\overline{M}$. One can directly see that qubits 1 and 2 must belong to different parts of the bipartition, otherwise the bipartition is only 1BP. We assume that $1 \in M$ and $2 \in \overline{M}$. Furthermore, among the remaining qubits $\{3, ..., n+1\}$, there must be at least one belonging to M and at least one belonging to $\overline{M}$. Otherwise, the bipartition would be a one-vs.-n-bipartition, which can never be a 2BP bipartition. Since $n \geq 5$ there must be either two qubits from the set $\{3, ..., n+1\}$ in M or two qubits from the set $\{3, ..., n+1\}$ in $\overline{M}$. Let us assume that the two qubits $i, j \in \{3, ..., n+1\}$ belong to M.

Now we apply local complementation on qubit 3 and then on qubit i [see Fig. 6.2 c)]. If $i = 3$ this changes nothing. Otherwise, the graph is transformed so that qubits 3 and i are interchanged, hence the qubit i is afterwards the "central" qubit. Qubit j is now only connected to the qubit i, and both qubits belong to the same part of the bipartition.

This, however, is exactly the situation as described in Lemma 56. As in the proof of Lemma 56, we can decouple the qubits i and j, and the remaining two states $\varrho^{\pm}$ are Y_N-graph-diagonal states of N qubits, which are PPT with respect to the given 2BP bipartition. By the induction hypothesis, these states are PPT mixtures with respect to 1BP bipartitions. Translating this backwards by inserting again the previously deleted connection finally proves the claim. □

Theorem 59. *A Y_N-graph-diagonal state is biseparable, if and only if it is a PPT mixture.*

Proof. The proof is essentially the same as that of Theorem 57. We only have to consider 1BP bipartitions according to Lemma 58 and for them PPT is necessary and sufficient. Note that for the Y_N graph there are only 1BP and 2BP bipartitions, 3BP bipartitions are not possible. □

As for the Y_5-state discussed after Theorem 57, one can simplify the search for PPT mixtures for the Y_N state by concentrating only on 1BP bipartitions. This makes it possible to determine separability for Y_N-graph-diagonal states for larger values of N, although the number of 1BP bipartitions still grows fast.

6.5.2 Biseparable decompositions for linear cluster states

For our second example of separability conditions for graph states of more than five particles, let us discuss the six-qubit linear cluster state mixed with white noise. Our goal is to show that the criteria used in this chapter allow estimates of separable regions in a simple way even for graph-diagonal states with many qubits, and the resulting estimate is quite accurate.

First, in a straightforward generalization of Lemma 49, many pairs of the form $\sigma = |000000\rangle\langle\cdot| + |ijklmn\rangle\langle\cdot|$ are separable, the exceptions are the 44 states $\chi_i = |000000\rangle\langle\cdot| + \eta_i$, with $\eta_1 = |1jk100\rangle\langle\cdot|$, $\eta_2 = |1jkl10\rangle\langle\cdot|$, $\eta_3 = |1jklm1\rangle\langle\cdot|$, $\eta_4 = |001jk1\rangle\langle\cdot|$, $\eta_5 = |01jkl1\rangle\langle\cdot|$ and $\eta_6 = |01jk10\rangle\langle\cdot|$. Furthermore, using the fact that the state from Eq. (6.14) is separable, one can directly find a biseparable decomposition of

$$\varrho(p) = p|Cl_6\rangle\langle Cl_6| + (1-p)\frac{1}{64} \qquad (6.24)$$

for $p = 11/43 \approx 0.256$. Since the state $\varrho(p)$ is known to be entangled for $p > 51/179 \approx 0.285$ (cf. Table 5.1 in Sec. 5), the real threshold cannot be much higher and this simple estimate already delivers a good approximation.

This method of constructing biseparable decompositions in the graph basis of linear cluster states can be generalized to an arbitrary number of qubits.

6.6 Conclusion

In conclusion, we have considered the problem of detecting genuine multiparticle entanglement in graph-diagonal states for four and five qubits and we have provided complete solutions for some cases. In addition, we showed how these results allow us to gain insight into this problem for larger numbers of qubits. Since our results deliver optimal criteria, they can be used to test the strength of other entanglement criteria.

In this chapter, we have made use of the graph formalism, which allowed us to find the presented solutions. A natural direction for future research would be to investigate the entanglement properties of other classes of states. Most promising candidates are states that exhibit a nice structure, such as invariance under permutations or under other kind of operations. Also, other kinds of entanglement, like states that are not fully separable (for first results see Ref. [121, 122]), are possible areas of investigation.

7 Multipartite Leggett models

Finally, we come back to the foundations of quantum mechanics and consider the class of hidden-variable models proposed by Leggett [28]. For the bipartite case, there exist some inequalities [59, 60, 123, 124], to one of which the reader was introduced in Sec. 2.4.3.

Very recently, there have been proposals of Leggett inequalities for the multipartite case. For example, Ref. [125] constructs a Leggett inequality for three qubits by taking Eq. (2.55) and replacing the observable A_i on the left-hand side by two observables $A_i C_i$ on the first and the third qubit, while the right-hand side stays the same. They find that the resulting inequality is violated by the GHZ state. This inequality can also be violated by a state like $|\psi^-\rangle\langle\psi^-| \otimes |0\rangle\langle 0|$ with the observable $C_1 = C_2 = C_3 = Z$, since in this case, it reduces to Eq. (2.55).

Another preprint derives a different inequality for four qubits under the assumption that the reduced two-qubit state of Alice and Bob is pure and that each single-qubit state of Alice and each single-qubit state of Bob is pure [126].

In the following, similarly to multipartite entanglement or multipartite non-locality [54, 57], we pursue different approaches as we take into account all bipartitons and, moreover, introduce different Leggett models in a way similar to the distinction of full separability and biseparability in the case of entanglement.

7.1 Leggett's assumption on one qubit

First, we introduce models that impose assumptions on single-qubit states. Analogously with full separability, we introduce the following notion.

Definition 60. *Any probability distribution $p(\alpha, \beta, \gamma | A, B, C)$ on the measurement outcomes α, β, γ when measuring the observables A, B and C describes a **full Leggett model** if it can be written as*

$$p(\alpha, \beta, \gamma | A, B, C) = \int d\lambda \varrho(\lambda) p_\lambda(\alpha, \beta, \gamma | A, B, C), \tag{7.1}$$

where for each λ with $\varrho(\lambda) \neq 0$, the distribution $p_\lambda(\alpha, \beta, \gamma | A, B, C)$ fulfills the following two conditions:

1. Its three marginals can be written as

$$p_\lambda(\alpha|A,B,C) = \vec{\mu}\vec{a},$$
$$p_\lambda(\beta|A,B,C) = \vec{\nu}\vec{b},$$
$$p_\lambda(\gamma|A,B,C) = \vec{\omega}\vec{c}. \tag{7.2}$$

Here, $\vec{\mu}, \vec{\nu}, \vec{\omega} \in \mathbb{R}^3$ are unit vectors and the unit vectors $\vec{a}, \vec{b}, \vec{c} \in \mathbb{R}^3$ are given by the observables A, B and C.

2. The distribution $p(\alpha,\beta,\gamma|A,B,C)$ is no-signalling with respect to Alice, i.e. $p(\alpha|A,B,C) = p(\alpha|A)$ and the same for Bob and Charlie and $p(\alpha,\beta|A,B,C) = p(\alpha,\beta|A,B)$ and the same for the two other cases.

Note that no-signalling of one-qubit marginals such as $p_\lambda(\alpha|A,B,C)$ is implied by condition (i). However, no-signalling conditions of probability distributions on two parties, such as $p(\alpha,\beta|A,B,C) = p(\alpha,\beta|A,B)$, is not implied by (i). For this reason, condition (ii) is really needed. Moreover, we note that, in general, measurements on a biseparable state such as $|\psi\rangle = |\psi_{AB}\rangle \otimes |\psi_C\rangle$ are enough to produce probabilities that do not obey condition (i).

As we would like the models that we consider to include a large class of probability distributions, we can relax condition (i) and, analogously to biseparability and the class of models that obey the Svetlichny inequality [57], introduce the notion of hybrid Leggett models.

Definition 61. Any probability distribution $p(\alpha,\beta,\gamma|A,B,C)$ on the measurement outcomes α, β, γ when measuring the observables A, B and C belongs to a **hybrid single-party Leggett model** if it obeys the following conditions:

(i) It can be written as

$$\begin{aligned}p(\alpha,\beta,\gamma|A,B,C) &= q_1 p_{\langle A\rangle BC}(\alpha,\beta,\gamma|A,B,C) \\&+ q_2 p_{A\langle B\rangle C}(\alpha,\beta,\gamma|A,B,C) \\&+ q_3 p_{AB\langle C\rangle}(\alpha,\beta,\gamma|A,B,C),\end{aligned} \tag{7.3}$$

7.1 Leggett's assumption on one qubit

with $\sum_i q_i = 1$ and $q_i \geq 0$, where

$$p_{\langle A \rangle BC}(\alpha, \beta, \gamma | A, B, C) = \int d\lambda \varrho(\lambda) p_\lambda^A(\alpha, \beta, \gamma | A, B, C), \tag{7.4}$$

with $\langle A \rangle_\lambda = \vec{\mu}\vec{a}$ and

$$p_{A\langle B\rangle C}(\alpha, \beta, \gamma | A, B, C) = \int d\lambda \varrho(\lambda) p_\lambda^B(\alpha, \beta, \gamma | A, B, C), \tag{7.5}$$

with $\langle B \rangle_\lambda = \vec{\nu}\vec{b}$ and

$$p_{AB\langle C\rangle}(\alpha, \beta, \gamma | A, B, C) = \int d\lambda \varrho(\lambda) p_\lambda^C(\alpha, \beta, \gamma | A, B, C), \tag{7.6}$$

with $\langle C \rangle_\lambda = \vec{\omega}\vec{c}$. \quad (7.7)

As before, $\vec{\mu}, \vec{\nu}, \vec{\omega} \in \mathbb{R}^3$ are unit vectors. Also, the unit vectors $\vec{a}, \vec{b}, \vec{c} \in \mathbb{R}^3$ are given by the observables A, B and C.

(ii) The distribution is no-signalling with respect to any bipartition as in Definition 60 (ii).

Probability distributions of this kind include all distributions of Definition 60. Note that, in the case of four parties, there are not only bipartitions with one party on one and two parties on the other side, but also 2|2-bipartitions. The two present notions of Leggett models therefore have to be modified for the case of a larger number of parties. We will do so in Sec. 7.2. Before that, we first derive an inequality which holds for models as in Definition 61 in the following section.

7.1.1 An inequality for hybrid Leggett models with assumptions on a single party

Let us consider a setting of three parties. There are two observables A and A' for Alice, B and B' for Bob and C and C' for Charlie. Since there are 8 ways to combine these six observables to one that acts on all three qubits, there are eight probability distributions of the kind $p_\lambda(\alpha, \beta, \gamma | A, B, C)$, but with respect to different observables. Moreover, using no-signalling, they can be written as

$$p_\lambda(\alpha, \beta, \gamma | A, B, C)$$
$$= \frac{1}{8}\left(\alpha\langle A \rangle + \beta\langle B \rangle + \gamma\langle C \rangle + \alpha\beta\langle AB \rangle + \beta\gamma\langle BC \rangle + \alpha\gamma\langle AC \rangle + \alpha\beta\gamma\langle ABC \rangle\right). \tag{7.8}$$

Adding them with appropriately chosen values for α, β and γ and using the positivity the distributions and therefore of their sum, one arrives at

$$|\langle ABC \rangle_\lambda + \langle ABC' \rangle_\lambda + \langle AB'C \rangle_\lambda + \langle A'BC \rangle_\lambda$$
$$+ \langle A'B'C \rangle_\lambda + \langle A'BC' \rangle_\lambda + \langle AB'C' \rangle_\lambda + \langle A'B'C' \rangle_\lambda|$$
$$\leq 8 - 4|\langle A \rangle_\lambda - \langle A' \rangle_\lambda|. \tag{7.9}$$

7 Multipartite Leggett models

Note that, in principle, such an upper bound of the left-hand side has to be derived for each of the three terms in Eq. (7.3). However, since the left-hand side of Eq. (7.9) is invariant under permutation of qubits, one immediately knows that the right-hand side is also an upper bound when replacing A by B and A' by B' or by C and C'. These three possible upper bounds can be used for the three terms in Eq. (7.3). Thus, let us consider distribution $p_\lambda^A(a,b,c|A,B,C)$ [cf. Eq. (7.4)], where the form of the one-qubit expectation values for Alice is fixed.

Integration $\int d\lambda \varrho(\lambda)$ of both sides of Eq. (7.9) and the use of $|\int .| \leq \int |.|$ then results in

$$|\langle ABC \rangle + \langle ABC' \rangle + \langle AB'C \rangle + \langle A'BC \rangle$$
$$+ \langle A'B'C \rangle + \langle A'BC' \rangle + \langle AB'C' \rangle + \langle A'B'C' \rangle|$$
$$\leq 8 - 4 \int d\lambda \varrho(\lambda) |\langle A \rangle_\lambda - \langle A' \rangle_\lambda| \qquad (7.10)$$

Now, we pass to a larger number of different settings for each observables. In other words, we do not only consider expectation values like $\langle ABC \rangle$, $\langle ABC' \rangle$ etc., but instead $\langle A_i B_i C_i \rangle$, $\langle A_i B_i C_i' \rangle$ etc., where i can take values from 1 to 3. For brevity, let us denote the left-hand side of Eq. (7.10) for the ith setting of observables by $\mathcal{L}_i = |\langle A_i B_i C_i \rangle + \langle A_i B_i C_i' \rangle + \ldots |$. Using Eq. (7.4), we then obtain

$$\frac{1}{m} \sum_{i=1}^m \mathcal{L}_i \leq 8 - 4 \frac{1}{m} \sum_{i=1}^m \int d\lambda \varrho(\lambda) |(\vec{a}_i - \vec{a}_i')\vec{\mu}| \, . \qquad (7.11)$$

Choosing $m = 3$ and the settings of Table 7.1.1, we can write

$$\frac{1}{3} \sum_{i=1}^3 \mathcal{L}_i \leq 8 - \sin(\varphi/2) \frac{8}{3} \sum_{i=1}^3 \int d\lambda \varrho(\lambda) |\vec{v}_i \vec{\mu}| \, . \qquad (7.12)$$

Here, the unit vectors $\vec{v}_i$ are given by $\vec{v}_1 = \vec{e}_y$, $\vec{v}_2 = \vec{u}^\perp$ and $\vec{v}_3 = \vec{e}_z$. Using the theorem of Ref. [127], one can see that $\sum_{i=1}^m |\vec{v}_i \vec{\mu}| \geq \frac{\sqrt{3}}{2}$. This bound is actually obtained for $\vec{\mu} = \vec{e}_x$. When we also plug in the GHZ state and the settings of Table 7.1.1 on the left-hand side, we obtain

$$8 \cos^3(\varphi/2) \leq 8 - \sqrt{3} \frac{4}{3} \sin(\varphi/2) \, . \qquad (7.13)$$

As mentioned before, the expression on the left-hand side of Eq. (7.10) is invariant under any permutations. When, e.g., permuting parties A and B, the inequality still holds, but this time for probability distributions $p_\lambda^B(\alpha, \beta, \gamma|A, B, C)$ and after replacing $\vec{a}_i$ and $\vec{a}_i'$ by $\vec{b}_i$ and $\vec{b}_i'$ on the right-hand side. In this way, it is possible to find upper bounds for all three terms in Eq. (7.3). Since the settings of Table 7.1.1 are the same for each party, the upper bound is always the same and Eq. (7.13) therefore holds for all hybrid single-party Leggett models.

The maximal value of the left-hand side Eq. (7.13) is reached for $\varphi \approx 11.26°$ and equals 7.88, while the right-hand side is 7.77. The left- and right-hand side are shown in Figure 7.1 b).

7.1 Leggett's assumption on one qubit

i	$\vec{a}_i, \vec{b}_i, \vec{c}_i$	$\vec{a}'_i, \vec{b}'_i, \vec{c}'_i$
1	$\cos(\varphi/2)\vec{e}_x + \sin(\varphi/2)\vec{e}_y$	$\cos(\varphi/2)\vec{e}_x - \sin(\varphi/2)\vec{e}_y$
2	$\cos(\varphi/2)\vec{u} + \sin(\varphi/2)\vec{u}^\perp$	$\cos(\varphi/2)\vec{u} - \sin(\varphi/2)\vec{u}^\perp$
3	$\cos(\varphi/2)\vec{u} + \sin(\varphi/2)\vec{e}_z$	$\cos(\varphi/2)\vec{u} - \sin(\varphi/2)\vec{e}_z$

Alice, Bob and Charlie each choose the same six settings. Here, $\vec{u} = -\frac{1}{2}\vec{e}_x - \frac{\sqrt{3}}{2}\vec{e}_y$ and $\vec{u}^\perp = \frac{\sqrt{3}}{2}\vec{e}_x - \frac{1}{2}\vec{e}_y$.

Note that the above construction in principle also works for others state under certain conditions. For a state $|\psi\rangle$, one only needs a triple A, B, C of observables, such that $A \otimes B \otimes C|\psi\rangle = |\psi\rangle$ and another triple $\widetilde{A}$, $\widetilde{B}$ and $\widetilde{C}$ with $\widetilde{A} \otimes \widetilde{B} \otimes \widetilde{C}|\psi\rangle = |\psi\rangle$, where Alice's two observables are orthogonal, i.e. obey $\text{Tr}(A\widetilde{A}) = 0$ and the same holds for Bob and Charlie. In this case, one can choose the measured observables as in Table 7.1.1, with $\vec{e}_x$ being replaced by the vector associated to A, $\vec{u}$ replaced by the vector associated with $\widetilde{A}$ and the three contributions containing $\sin(\varphi/2)$ must be linearly independent.

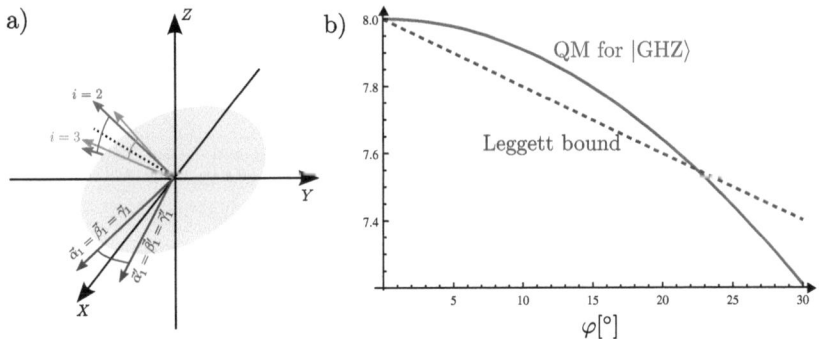

Figure 7.1: By choosing the six settings visualized in a) for each party, one obtains a Leggett bound that is shown as a dashed line in b). The quantum mechanical value for the GHZ state and the given observables is depicted by a solid line. The maximal violation is reached for $\varphi \approx 11.26°$.

7.2 An assumption on two qubits

For three qubits, each bipartition is made up of one party on one side and two parties on the other side. Thus, for every possible bipartition, one can impose Leggett's original assumption on the one-qubit side. For a larger number of parties, however, there are other bipartitions, such as the bipartition $AB|CD$ in the four-party case. Although it is not possible to apply the original assumption of Leggett, one can naturally extend the model by postulating that reduced two-party states behave as pure qubit states.

Before we can define this class of models properly, we need to find a general form for two-party expectation values that stem from a pure two-qubit state. This can be done using the Schmidt decomposition (cf. Lemma 2 in Sec. 2.1.1). In this case, it states that any pure two-qubit state is LU-equivalent with the state

$$|\psi\rangle = \cos(\phi)|00\rangle + \sin(\phi)|11\rangle \tag{7.14}$$

for some value of ϕ. In terms of Pauli matrices, this state can be written as

$$|\psi\rangle\langle\psi| = \frac{1}{2}\cos(\phi)\sin(\phi)X_1X_2 - \frac{1}{2}\cos(\phi)\sin(\phi)Y_1Y_2 + \frac{1}{4}Z_1Z_2 + \frac{1}{2}\mathbb{1}_1\mathbb{1}_2$$
$$+ \frac{1}{4}\left[\cos^2(\phi) - \sin^2(\phi)\right](Z_1\mathbb{1}_2 + \mathbb{1}_1Z_2) \tag{7.15}$$

$$= \frac{1}{4}\sum_{i,j=0}^{3} \psi_{i,j}\, \sigma_i \otimes \sigma_j\,. \tag{7.16}$$

In the last line, we wrote $|\psi\rangle$ in terms of the Pauli matrices σ_1, σ_2 and σ_3 and the identity σ_0. The expectation value with respect to some two-qubit observable $A \otimes B$ with $A = \sum_{i=1}^{3} a_i\sigma_i$ and $B = \sum_{j=1}^{3} b_j\sigma_j$ is given by

$$|\psi\rangle\langle\psi| = \sum_{i,j=1}^{3} a_i\psi_{i,j}b_j\,. \tag{7.17}$$

Thus, the only relevant part of the matrix $(\psi_{i,j})$ are the indices from 1 to 3.

As $|\psi\rangle$ is any two-qubit state up to local unitaries, we still need to take them into account. As every unitary U with negative determinant can be written as $e^{i\varphi}\widetilde{U}$, where $\widetilde{U}$ has positive determinant, and also $U|\psi\rangle\langle\psi|U^\dagger = \widetilde{U}|\psi\rangle\langle\psi|\widetilde{U}^\dagger$, it is sufficient to consider local unitaries with positive determinant. Furthermore, any unitary transformation (with determinant 1) on $|\psi\rangle\langle\psi|$ corresponds to an orthogonal three-by-three matrix (with determinant 1) acting on $\psi_{i,j}$ [128]. Thus, the expectation value of any pure two-qubit state with respect to any observable $A \otimes B$ can be written as

$$\text{Tr}(A \otimes B\, U_A \otimes U_B|\psi\rangle\langle\psi|U_A^\dagger \otimes U_B^\dagger) = \vec{a}O^{-1}\psi\widetilde{O}\vec{b}\,, \tag{7.18}$$

7.2 An assumption on two qubits

where ψ is a diagonal matrix with diagonal $(2\cos(\phi)\sin(\phi), -2\cos(\phi)\sin(\phi), 1)$ and O and $\tilde{O}$ are orthogonal matrices with determinant 1. All three matrices are determined by the state of the system, while the vectors $\vec{a}, \vec{b} \in \mathbb{R}^3$ are determined by the observables.

Thus, we can now extend the Leggett model on two parties as follows.

Definition 62. *Any probability distribution $p(\alpha, \beta, \gamma, \delta | A, B, C, D)$ on the measurement outcomes α, β, γ, δ when measuring the observables A, B, C and D describes a* **hybrid Leggett model** *if it obeys the following conditions:*

(i) The probability distribution can be written as

$$p(\alpha, \beta, \gamma, \delta | A, B, C, D) = q_1 p_{\langle A \rangle BCD} + q_2 p_{A\langle B\rangle CD} + q_3 p_{AB\langle C\rangle D} + q_4 p_{ABC\langle D\rangle}$$
$$q_5 p_{AB\langle CD\rangle} + q_6 p_{A\langle BC\rangle D} + q_7 p_{\langle AB\rangle CD} + q_8 p_{\langle A\rangle B\langle C\rangle D}$$
$$q_9 p_{A\langle B\rangle C\langle D\rangle} + q_{10} p_{\langle A\rangle BC\langle D\rangle}, \quad (7.19)$$

where the dependencies have been left out, with $\sum_i q_i = 1$ and $q_i \geq 0$. Here, we impose an assumption on one party on the first four terms, namely that they are of the type

$$p_{\langle A\rangle BCD}(\alpha, \beta, \gamma, \delta | A, B, C, D) = \int d\lambda \varrho(\lambda) p_\lambda^A(\alpha, \beta, \gamma, \delta | A, B, C, D) \quad (7.20)$$

with $\langle A \rangle_\lambda = \vec{\mu}\vec{a}$, where $\vec{\mu}$ and $\vec{a}$ are unit vectors in $\mathbb{R}^3$. Moreover, on the last six terms we impose a two-party assumption, namely that they are of the form

$$p_{\langle AB\rangle CD}(\alpha, \beta, \gamma, \delta | A, B, C, D) = \int d\lambda \varrho(\lambda) p_\lambda^{AB}(\alpha, \beta, \gamma, \delta | A, B, C, D) \quad (7.21)$$

with $\langle AB\rangle_\lambda = \vec{a} O^{-1} \psi \tilde{O} \vec{b}$, where $\vec{a}, \vec{b} \in \mathbb{R}^3$ are unit vectors, $O, \tilde{O} \in SO(3)$ and

$$\psi = \begin{pmatrix} 2\cos(\phi)\sin(\phi) & 0 & 0 \\ 0 & -2\cos(\phi)\sin(\phi) & 0 \\ 0 & 0 & 1 \end{pmatrix} \quad (7.22)$$

(ii) The distribution is no-signalling with respect to any bipartition as in Definition 60 (ii).

To derive an inequality for such a model, one could start with two observables for each of the four parties and consider the sum of all possible combinations,

$$|\langle ABCD\rangle + \langle ABCD'\rangle + \langle ABC'D\rangle + \cdots + \langle A'B'C'D\rangle + \langle A'B'C'D'\rangle|. \quad (7.23)$$

Since this is symmetric, it is enough to find an upper bound for the two bipartitions $A|BCD$ and $AB|CD$, i.e. for the probability distributions $p_{\langle A\rangle BCD}$ and $p_{\langle AB\rangle CD}$. While for $A|BCD$, an upper bound similar as in Eq. (7.10) can be used, for $AB|CD$, an upper bound like

$$|\langle ABCD\rangle + \langle ABCD'\rangle + \langle ABC'D\rangle + \cdots + \langle A'B'C'D\rangle + \langle A'B'C'D'\rangle|$$
$$\leq 16 - 4|\langle AB\rangle - \langle A'B'\rangle| - 4|\langle A'B\rangle - \langle AB'\rangle| \qquad (7.24)$$

can be used. Introducing several settings and summing over them, one eventually has to bound a term of the form $\sum_{i=1}^{m} |\vec{e}_i O^{-1} \psi \widetilde{O} \vec{f}_i|$, where the vectors $\vec{e}_i$ and $\vec{f}_i$ are given by the chosen settings. This can be done using operator norms. At the same time, the inequality should be violated for some quantum mechanical state. So far, however, an appropriate inequality that is violated by quantum mechanics remains to be found.

7.3 Conclusion

In this chapter, we have presented several ways to extend Leggett's assumption from Ref. [28] to the multipartite case. We have introduced a rather large and general class of multipartite Leggett models, in that we do not only consider one fixed bipartition, but convex combinations of probability distributions of different bipartitions. For one of the presented multipartite Leggett models, namely one for three parties with assumptions on single-party marginals, we have introduced an inequality which is violated by quantum mechanics.

As this is work in progress, a natural future direction would be to search for inequalities to test models that make assumptions on two-party expectation values or involve a larger number of parties.

8 Conclusion

In this thesis, different aspects of multipartite entanglement and non-locality have been investigated. Mainly, we focused on its characterization and detection.

In Sec. 3, we started with an analysis of the statistical error and the statistical significance in experiments that test for quantum correlations. It showed that the maximal possible violation is not a meaningful measure for the statistical strength of a non-locality test. This was due to the fact that the Mermin inequality, which only allows for a comparatively small violation, resulted in a higher statistical significance than the Ardehali inequality. For experiments that aim at the verification of non-locality or entanglement and which achieve a certain fidelity with respect to the target state, this finding motivates the use of inequalities which contain stabilizing operators of the target state (cf., e.g. Ref. [72]). Moreover, we could show in the case of white noise that the fidelity that needs to be reached decreases exponentially fast with a growing number of qubits.

Our findings can therefore be employed for developing statistically strong entanglement tests. Besides the mentioned exponential decrease in the required fidelity for white noise, the fact that the measured count numbers are decreasing for a higher number of photons means that the statistical error plays a more and more important role. Note that also for other implementations the count numbers per measurement setting usually decrease with a growing number of particles, as the number of required measurement settings for, say, a full tomography increases and the time spent on measuring can, for practical reasons, not be increased equally.

Then, we presented a criterion for genuine multipartite entanglement and the idea behind it, along with its most important properties, in Sec. 4. Most importantly, we introduced an entanglement monotone for genuine multipartite entanglement and we showed that our criterion is necessary and sufficient for three-qubit permutation-invariant states. We also applied it to several example states to illustrate its performance, to present an analytical witness for the W_3 state and to show its application in the case of partial information.

In order to obtain analytical tests for entanglement, we applied the criterion to graph states in Sec. 5. This resulted in a list of witnesses for all LU-equivalence classes for graph states of up to six qubits and in general construction methods for two different classes of entanglement witnesses. These witnesses were much stronger than the usual projector witness. Indeed their white noise tolerance approached one in many cases for an increasing number of qubits. Also, we evaluated the presented entanglement monotone on graph states.

Finally, in Sec. 6, we used the graph state witnesses constructed before and the approach introduced in Sec. 4 to construct necessary and sufficient conditions for entanglement in some classes of graph-diagonal states. These conditions were strongly connected to the entanglement criterion introduced in Sec. 4 and, therefore, our approach resulted in more classes of states for which this criterion is necessary and sufficient.

We believe that both the presented criterion — including its applications, such as the graph state witnesses — and the entanglement monotone will be useful for different tasks. First, the approach that lead to our criterion can be applied to other bipartite criteria, such as the criterion of symmetric extensions by Doherty [95, 96]. Second, the presented form of the criterion is an easy way to verify entanglement of an arbitrary state and can be applied to quickly check for entanglement in experimental states. And finally, the plethora of graph state witnesses presented provides much stronger entanglement tests than known so far, with only one experimental setting to be measured in addition to make the projector witnesses stronger. Besides this, our entanglement monotone provides an easy way to quantify genuine multipartite entanglement present in interesting states, such as thermal states in spin chain models.

To conclude this thesis, we presented some generalizations of the Leggett model of two parties to the multiparty case [28]. We provided an inequality which shows that quantum mechanics is incompatible with a version of such a model that is a convex combination of models which impose an assumption on the one-particle marginals. There are still many open questions and many possible ways to continue these investigations, such as inequalities for other versions of this model, in particular the ones that impose assumptions on two-party expectation values, and an extension to an arbitrary number of qubits. We hope that these investigations help to characterize and understand the properties of quantum mechanics better.

Bibliography

[1] Einstein, A., Podolsky, B., and Rosen, N. *Phys. Rev.* **47**, 777 (1935).

[2] Bell, J. S. *Physics* **1**, 195 (1964).

[3] Schrödinger, E. *Die Naturwissenschaften* **23**, 807–812; 823–828; 844–849 (1935).

[4] Werner, R. F. *Phys. Rev. A* **40**, 4277 (1989).

[5] Ekert, A. K. *Phys. Rev. Lett.* **67**, 661 (1991).

[6] Bennett, C., Brassard, G., Crpeau, C., Jozsa, R., Peres, A., and Wootters, W. *Phys. Rev. Lett.* **70**, 1895 (1993).

[7] Raussendorf, R. and Briegel, H. *Phys. Rev. Lett.* **86**, 5188 (2001).

[8] Briegel, H. J., Browne, D. E., Dür, W., Raussendorf, R., and Van den Nest, M. *Nat. Phys.* **5**, 19 (2009).

[9] Shor, P. *SIAM J. Comp.* **26**, 1474 (1997).

[10] Grover, L. *Phys. Rev. Lett. 79* **79**, 325 (1997).

[11] Vidal, G. *Phys. Rev. Lett.* **91**, 147902 (2003).

[12] Ollivier, H. and Zurek, W. H. *Phys. Rev. Lett.* **88**, 017901 (2001).

[13] Datta, A., Shaji, A., and Caves, C. M. *Phys. Rev. Lett.* **100**, 050502 (2008).

[14] Giovannetti, V., Lloyd, S., and Maccone, L. *Nature* **306**, 1330 (2004).

[15] Collins, D., Gisin, N., Popescu, S., Roberts, D., and Scarani, V. *Phys. Rev. Lett.* **88**, 170405 (2002).

[16] Gühne, O. and Seevinck, M. *New Journal of Physics* **12**, 053002 (2010).

[17] Huber, M., Mintert, F., Gabriel, A., and Hiesmayr, B. C. *Phys. Rev. Lett.* **104**, 210501 (2010).

[18] Horodecki, R., Horodecki, P., Horodecki, M., and Horodecki, K. *Rev. Mod. Phys.* **81**, 865–942 (2009).

[19] Gühne, O. and Tóth, G. *Physics Reports* **474**, 1 – 75 (2009).

[20] Dür, W., Vidal, G., and Cirac, J. I. *Phys. Rev. A* **62**, 062314 (2000).

[21] Acín, A., Andrianov, A., Costa, L., Jané, E., Latorre, J. I., and Tarrach, R. *Phys. Rev. Lett.* **85**, 1560–1563 (2000).

[22] Verstraete, F., Dehaene, J., De Moor, B., and Verschelde, H. *Phys. Rev. A* **65**, 052112 (2002).

[23] Lamata, L., León, J., Salgado, D., and Solano, E. *Phys. Rev. A* **74**, 052336 (2006).

[24] Monz, T., Schindler, P., Barreiro, J. T., Chwalla, M., Nigg, D., Coish, W. A., Harlander, M., Hänsel, W., Hennrich, M., and Blatt, R. *Phys. Rev. Lett.* **106**, 130506 (2011).

[25] Gao, W.-B., Lu, C.-Y., Yao, X.-C., Xu, P., Gühne, O., Goebel, A., Chen, Y.-A., Peng, C.-Z., Chen, Z.-B., and Pan, J.-W. *Nature Physics* **6**, 331 (2010).

[26] Neeley, M., Bialczak, R. C., Lenander, M., Lucero, E., Mariantoni, M., O'Connell, A. D., Sank, D., Wang, H., Weides, M., Wenner, J., Ying, Y., Yamamoto, T., Cleland, A. N., and Martinis, J. M. *Nature* **467**, 570 (2010).

[27] DiCarlo, L., Reed, M. D., Sun, L., Johnson, B. R., Chow, J. M., Gambetta, J. M., Frunzio, L., Girvin, S. M., Devoret, M. H., and Schoelkopf, R. J. *Nature* **467**, 574 (10).

[28] Leggett, A. *Foundations of Physics* **33**, 1469–1493 (2003).

[29] Nielsen, M. A. and Chuang, I. L. *Quantum computation and quantum information*. Cambridge University Press, (2000).

[30] Peres, A. *Phys. Rev. Lett.* **77**, 1413–1415 (1996).

[31] Horodecki, M., Horodecki, P., and Horodecki, R. *Physics Letters A* **223**, 1 – 8 (1996).

[32] Terhal, B. M. *Physics Letters A* **271**, 319 – 326 (2000).

[33] Lewenstein, M., Kraus, B., Cirac, J. I., and Horodecki, P. *Phys. Rev. A* **62**, 052310 (2000).

[34] Bennett, C. H., DiVincenzo, D. P., Fuchs, C. A., Mor, T., Rains, E., Shor, P. W., Smolin, J. A., and Wootters, W. K. *Phys. Rev. A* **59**, 1070–1091 (1999).

[35] Horodecki, M. *Quant. Inf. Comp.* **1**, 3 (2001).

[36] Plenio, M. B. and Virmani, S. *Quant. Inf. Comp.* **7**, 1 (2007).

[37] Bennett, C. H., DiVincenzo, D. P., Smolin, J. A., and Wootters, W. K. *Phys. Rev. A* **54**, 3824–3851 (1996).

Bibliography

[38] Vidal, G. and Werner, R. F. *Phys. Rev. A* **65**, 032314 (2002).

[39] Hill, S. and Wootters, W. K. *Phys. Rev. Lett.* **78**, 5022–5025 (1997).

[40] Rungta, P., Bužek, V., Caves, C. M., Hillery, M., and Milburn, G. J. *Phys. Rev. A* **64**, 042315 (2001).

[41] Wootters, W. K. *Phys. Rev. Lett.* **80**, 2245–2248 (1998).

[42] Shimony, A. *Annals of the New York Academy of Sciences* **755**, 675–679 (1995).

[43] Barnum, H. and Linden, N. *Journal of Physics A: Mathematical and General* **34**, 6787 (2001).

[44] Wei, T.-C. and Goldbart, P. M. *Phys. Rev. A* **68**, 042307 (2003).

[45] Vedral, V. and Plenio, M. B. *Phys. Rev. A* **57**, 1619–1633 (1998).

[46] Vidal, G. and Tarrach, R. *Phys. Rev. A* **59**, 141–155 (1999).

[47] Steiner, M. *Phys. Rev. A* **67**, 054305 (2003).

[48] Coffman, V., Kundu, J., and Wootters, W. K. *Phys. Rev. A* **61**, 052306 (2000).

[49] Lohmayer, R., Osterloh, A., Siewert, J., and Uhlmann, A. *Phys. Rev. Lett.* **97**, 260502 (2006).

[50] Osterloh, A. and Siewert, J. *Phys. Rev. A* **72**, 012337 (2005).

[51] Osterloh, A. and Siewert, J. *Int. J. Quantum Inf.* **4**, 531 (2006).

[52] Clauser, J. F., Horne, M. A., Shimony, A., and Holt, R. A. *Phys. Rev. Lett.* **23**, 880–884 (1969).

[53] Barrett, J. *Phys. Rev. A* **65**, 042302 (2002).

[54] Mermin, N. D. *Phys. Rev. Lett.* **65**, 1838–1840 (1990).

[55] Greenberger, D. M., Horne, M. A., and Zeilinger, A. *Bell's theorem, quantum theory and conceptions of the universe.* Kluwer, Dordrecht, (1989).

[56] Ardehali, M. *Phys. Rev. A* **46**, 5375–5378 (1992).

[57] Svetlichny, G. *Phys. Rev. D* **35**, 3066–3069 (1987).

[58] Mitchell, P., Popescu, S., and Roberts, D. *Phys. Rev. A* **70**, 060101 (2004).

[59] Gröblacher, S., Paterek, T., Kaltenbaek, R., Brukner, v., Żukowski, M., Aspelmeyer, M., and Zeilinger, A. *Nature* **446**, 871 – 875 (2007).

[60] Branciard, C., Brunner, N., Gisin, N., Kurtsiefer, C., Lamas-Linares, A., Ling, A., and Scarani, V. *Nature Physics* **4**, 681 – 685 (2008).

[61] Schlingemann, D. and Werner, R. F. *Phys. Rev. A* **65**, 012308 (2001).

[62] Grassl, M., Klappenecker, A., and Rötteler, M. In *Proceedings 2002 IEEE International Symposium on Information Theory*, (2002). Lausanne, Switzerland.

[63] Chen, K. and Lo, H.-K. *Quant. Inf. Comp.* **7**, 689 (2007).

[64] Hein, M., Dür, W., Eiser, J., Raussendorf, R., Van den Nest, M., and Briegel, H. J. In *Proceedings of the International School of Physics "Enrico Fermi" on "Quantum Computers, Algorithms and Chaos"*, (2006).

[65] Kiesel, N., Schmid, C., Weber, U., Tóth, G., Gühne, O., Ursin, R., and Weinfurter, H. *Phys. Rev. Lett.* **95**, 210502 (2005).

[66] Lu, C.-Y., Zhou, X.-Q., Gühne, O., Gao, W.-B., Zhang, J., Yuan, Z.-S., Goebel, A., Yang, T., and Pan, J.-W. *Nat. Phys.* **3**, 91 (2007).

[67] Ceccarelli, R., Vallone, G., De Martini, F., Mataloni, P., and Cabello, A. *Phys. Rev. Lett.* **103**, 160401 (2009).

[68] Gao, W.-B., Yao, X.-C., Xu, P., Lu, H., Gühne, O., Cabello, A., Lu, C.-Y., Yang, T., Chen, Z.-B., and Pan, J.-W. *Phys. Rev. A* **82**, 042334 (2010).

[69] Wunderlich, H., Vallone, G., Mataloni, P., and Plenio, M. B. *New Journal of Physics* **13**, 033033 (2011).

[70] Lee, S. M., Park, H. S., Cho, J., Kang, Y., Lee, J. Y., Kim, H., Lee, D.-H., and Choi, S.-K. arXiv:1105.5211.

[71] Yao, X.-C., Wang, T.-X., Xu, P., Lu, H., Pan, G.-S., Bao, X.-H., Peng, C.-Z., Lu, C.-Y., Chen, Y.-A., and Pan, J.-W. arXiv:1105.6318.

[72] Tóth, G. and Gühne, O. *Phys. Rev. Lett.* **94**, 060501 (2005).

[73] Tóth, G. and Gühne, O. *Phys. Rev. A* **72**, 022340 (2005).

[74] Briegel, H. J. and Raussendorf, R. *Phys. Rev. Lett.* **86**, 910–913 (2001).

[75] Jungnitsch, B., Moroder, T., and Gühne, O. *Phys. Rev. A* **84**, 032310 (2011).

[76] Aschauer, H., Dür, W., and Briegel, H.-J. *Phys. Rev. A* **71**, 012319 (2005).

[77] Hein, M., Eisert, J., and Briegel, H. J. *Phys. Rev. A* **69**, 062311 (2004).

Bibliography

[78] Van den Nest, M., Dehaene, J., and De Moor, B. *Phys. Rev. A* **69**, 022316 (2004).

[79] Cabello, A., López-Tarrida, A. J., Moreno, P., and Portillo, J. R. *Phys. Rev. A* **80**, 012102 (2009).

[80] Bourennane, M., Eibl, M., Kurtsiefer, C., Gaertner, S., Weinfurter, H., Gühne, O., Hyllus, P., Bruß, D., Lewenstein, M., and Sanpera, A. *Phys. Rev. Lett.* **92**, 087902 (2004).

[81] Vandenberghe, L. and Boyd, S. *SIAM Rev.* **38**, 49 (1996).

[82] Jungnitsch, B., Niekamp, S., Kleinmann, M., Gühne, O., Lu, H., Gao, W.-B., Chen, Y.-A., Chen, Z.-B., and Pan, J.-W. *Phys. Rev. Lett.* **104**, 210401 (2010).

[83] van Dam, W., Gill, R., and Grunwald, P. *Information Theory, IEEE Transactions on* **51**, 2812 – 2835 (2005).

[84] Acín, A., Gill, R., and Gisin, N. *Phys. Rev. Lett.* **95**, 210402 (2005).

[85] Altepeter, J. B., Jeffrey, E. R., Kwiat, P. G., Tanzilli, S., Gisin, N., and Acín, A. *Phys. Rev. Lett.* **95**, 033601 (2005).

[86] James, D. F. V., Kwiat, P. G., Munro, W. J., and White, A. G. *Phys. Rev. A* **64**, 052312 (2001).

[87] Chen, Y.-A., Zhang, A.-N., Zhao, Z., Zhou, X.-Q., and Pan, J.-W. *Phys. Rev. Lett.* **96**, 220504 (2006).

[88] Scarani, V., Acín, A., Schenck, E., and Aspelmeyer, M. *Phys. Rev. A* **71**, 042325 (2005).

[89] Cabello, A., Gühne, O., and Rodríguez, D. *Phys. Rev. A* **77**, 062106 (2008).

[90] Gühne, O. and Cabello, A. *Phys. Rev. A* **77**, 032108 (2008).

[91] Hofmann, H. F. *Phys. Rev. Lett.* **94**, 160504 (2005).

[92] Jungnitsch, B., Moroder, T., and Gühne, O. *Phys. Rev. Lett.* **106**, 190502 (2011).

[93] Piani, M. and Mora, C. E. *Phys. Rev. A* **75**, 012305 (2007).

[94] Tóth, G. and Gühne, O. *Phys. Rev. Lett.* **102**, 170503 (2009).

[95] Doherty, A. C., Parrilo, P. A., and Spedalieri, F. M. *Phys. Rev. Lett.* **88**, 187904 (2002).

[96] Doherty, A. C., Parrilo, P. A., and Spedalieri, F. M. *Phys. Rev. A* **69**, 022308 (2004).

[97] Löfberg, J. In *Proceedings of the CACSD Conference*, (2004).

[98] Sturm, J. F. *Opt. Meth. and Softw.* **11**, 625 (1999).

[99] Toh, K. C., Todd, M. J., and Tutuncu, R. H. *Opt. Meth. and Softw.* **11**, 545 (1999).

[100] Tutuncu, R. H., Toh, K. C., and Todd, M. J. *Opt. Meth. and Softw.* **95**, 189 (2003).

[101] Jungnitsch, B. (2011). PPTMixer: A tool to detect genuine multipartite entanglement, http://www.mathworks.com/matlabcentral/fileexchange/30968.

[102] Shchukin, E. and Vogel, W. *Phys. Rev. Lett.* **95**, 230502 (2005).

[103] Miranowicz, A., Piani, M., Horodecki, P., and Horodecki, R. *Phys. Rev. A* **80**, 052303 (2009).

[104] Häseler, H., Moroder, T., and Lütkenhaus, N. *Phys. Rev. A* **77**, 032303 (2008).

[105] Hyllus, P. and Eisert, J. *New Journal of Physics* **8**, 51 (2006).

[106] Plenio, M. B. *Phys. Rev. Lett.* **95**, 090503 (2005).

[107] Tokunaga, Y., Yamamoto, T., Koashi, M., and Imoto, N. *Phys. Rev. A* **74**, 020301 (2006).

[108] Huber, M., Erker, P., Schimpf, H., Gabriel, A., and Hiesmayr, B. *Phys. Rev. A* **83**, 040301 (2011).

[109] Gühne, O., Lu, C.-Y., Gao, W.-B., and Pan, J.-W. *Phys. Rev. A* **76**, 030305 (2007).

[110] Nielsen, M. A. *Phys. Rev. Lett.* **83**, 436–439 (1999).

[111] Gühne, O., Jungnitsch, B., Moroder, T., and Weinstein, Y. S. *Phys. Rev. A* **84**, 052319 (2011).

[112] Hein, M., Dür, W., and Briegel, H.-J. *Phys. Rev. A* **71**, 032350 (2005).

[113] Cavalcanti, D., Chaves, R., Aolita, L., Davidovich, L., and Acín, A. *Phys. Rev. Lett.* **103**, 030502 (2009).

[114] Weinstein, Y. S. *Phys. Rev. A* **80**, 022310 (2009).

[115] Horodecki, R. and Horodecki, M. *Phys. Rev. A* **54**, 1838–1843 (1996).

[116] Smolin, J. A. *Phys. Rev. A* **63**, 032306 (2001).

[117] Horodecki, P., Lewenstein, M., Vidal, G., and Cirac, I. *Phys. Rev. A* **62**, 032310 (2000).

[118] Benatti, F., Floreanini, R., and Piani, M. *Open Systems & Information Dynamics* **11**, 325–338 (2004).

[119] Piani, M. *Phys. Rev. A* **73**, 012345 (2006).

[120] Horodecki, M., Horodecki, P., and Horodecki, R. *Phys. Rev. Lett.* **80**, 5239–5242 (1998).

Bibliography

[121] Kay, A. *Journal of Physics A: Mathematical and Theoretical* **43**, 495301 (2010).

[122] Kay, A. *Phys. Rev. A* **83**, 020303 (2011).

[123] Paterek, T., Fedrizzi, A., Gröblacher, S., Jennewein, T., Żukowski, M., Aspelmeyer, M., and Zeilinger, A. *Phys. Rev. Lett.* **99**, 210406 (2007).

[124] Branciard, C., Ling, A., Gisin, N., Kurtsiefer, C., Lamas-Linares, A., and Scarani, V. *Phys. Rev. Lett.* **99**, 210407 (2007).

[125] Deng, D.-L., Wu, C., Chen, J.-L., and Oh, C. H. arXiv:1111.4119v1.

[126] Wieśniak, M. arXiv:1111.4410v1.

[127] Rai, A., Home, D., and Majumdar, A. S. *Phys. Rev. A* **84**, 052115 (2011).

[128] Hall, B. C. *Lie groups, Lie algebras and representations: An elementary introduction.* Springer, (2003).

Publications

(i) B. Jungnitsch and J. Evers, *Parametric and nonparametric magnetic response enhancement via electrically induced magnetic moments*, Phys. Rev. A **78**, 043817 (2008)

(ii) B. Jungnitsch, S. Niekamp, M. Kleinmann, O. Gühne, H. Lu, W.-B. Gao, Y.-A. Chen, Z.-B. Chen and Jian-Wei Pan, *Increasing the statistical significance of entanglement detection in experiments*, Phys. Rev. Lett. **104**, 210401 (2010)

(iii) B. Jungnitsch, T. Moroder and O. Gühne, *Taming multiparticle entanglement*, Phys. Rev. Lett. **106**, 190502 (2011)

(iv) B. Jungnitsch, *PPTMixer: A tool to detect genuine multipartite entanglement*, Matlab Central File Exchange, http://www.mathworks.com/matlabcentral/fileexchange/30968 (2001)

(v) B. Jungnitsch, T. Moroder and O. Gühne, *Entanglement Witnesses for Graph States: General Theory and Examples*, Phys. Rev. A **84**, 032310 (2011)

(vi) O. Gühne, B. Jungnitsch, T. Moroder und Y. S. Weinstein, *Multiparticle entanglement in graph-diagonal states*, Phys. Rev. A **84**, 052319 (2011)

The papers (ii), (iii), (v), (vi) and the program (iv) were created in the course of this thesis.

Acknowledgments

This thesis could not have been written without the support and help of many people who I would like to thank in the following.

- I would like to especially thank Prof. Dr. Otfried Gühne, who is now at the University of Siegen, for his permanent support and supervision, his valuable input and his optimism that was a constant source of motivation for me. Most importantly, I very much appreciated that he always had time for discussions and questions and the he enabled me to attend plenty of conferences and meet a lot of bright people.

- Moreover, I would like to thank Prof. Dr. Hans Briegel, who let me be a member of his active research group and the vibrant and enjoyable environment in Innsbruck. He supported me whenever I was in need of anything.

- Furthermore, the uncomplicated service by the administrative staff at IQOQI and the Institute for Theoretical Physics was a great help and provided the frame in which this thesis could be created.

- Also, my colleagues in Innsbruck and Siegen helped to create a very comfortable atmosphere at work and many of them became good friends. Among many others, I would like to thank Dr. Tobias Moroder for the close cooperation that I always enjoyed.

- I would also like to thank Julia Hupfeld. With her at my side, life was much more joyful and work was easier.

- Last but not least, I wish to thank my parents Birgit and Siegfried Jungnitsch, without whose support not only my PhD thesis, but also all the experiences I was able to make so far would have been impossible.

i want morebooks!

Buy your books fast and straightforward online - at one of world's fastest growing online book stores! Environmentally sound due to Print-on Demand technologies.

Buy your books online at
www.get-morebooks.com

Kaufen Sie Ihre Bücher schnell und unkompliziert online – auf einer der am schnellsten wachsenden Buchhandelsplattformen weltweit! Dank Print-On-Demand umwelt- und ressourcenschonend produziert.

Bücher schneller online kaufen
www.morebooks.de

VDM Verlagsservicegesellschaft mbH
Heinrich-Böcking-Str. 6-8   Telefon: +49 681 3720 174   info@vdm-vsg.de
D - 66121 Saarbrücken   Telefax: +49 681 3720 1749   www.vdm-vsg.de

Printed by Books on Demand GmbH, Norderstedt / Germany